WILLIAM LOBB, PLANT HUNTER FOR VEITCH AND MESSENGER OF THE BIG TREE

Manuscript submitted to *Huntia*, January, 1966, and set in galley before the journal was discontinued.

WILLIAM LOBB, PLANT HUNTER FOR VEITCH AND MESSENGER OF THE BIG TREE

BY

JOSEPH EWAN

UNIVERSITY OF CALIFORNIA PRESS

BERKELEY · LOS ANGELES · LONDON

1973

University of California Publications in Botany

Volume 67
Approved for publication April 27, 1973
Issued October 30, 1973

University of California Press
Berkeley and Los Angeles
California

—:—

University of California Press, Ltd.
London, England

ISBN: 0-520-09501-4
Library of Congress Card Number: 73-620118

printed in the united states of america

CONTENTS

Early Years 4

Lobb and the Big Tree 7

Later Years 10

William Lobb's Position in California Botany and Horticulture 14

Chronology 16

List of Horticultural Introductions and New Species 20

Finding List of Numbered California Collections 28

Acknowledgments 30

References 30

Index of Plant Names 35

For these travellers, whilst they enrich our museums with the different productions of the globe, whilst they increase our knowledge in geography, and point out new sources of commerce and new excitements to industry, too frequently fall victims to their zeal, before they have completed their enterprises, and have no opportunity of publishing an account of their labours and their discoveries. They have a claim to our acknowledgements; yet their names would sink into oblivion, unless some one could be found willing to pay them the tribute of writing their merited eulogy. It becomes then a duty in us to pay that respect to their memory, which we in vain regret they cannot now enjoy. [Karl Konig and John Sims, 1806.[1]]

It is not easy to realize the zeal of the Englishman for his garden and stove a century ago. While orchids, aroids, and choice exotics were the prizes of the well-to-do, the middle classes were gaining leisure and some affluence with the Victorian graces, cultivating their perennial borders, mulching the favorite conifer or tending the tender subjects in the glasshouse. Colonies were prospering and those living abroad, from diplomat to missionary and clerk, shared an interest in flowering plants. Curtis' *Botanical magazine* and other color-plate journals published novelties. Kew was flourishing and botanic gardens around the world kept her ranges and great Palm House abloom. Henry B. May, in writing of his memories in his *Seventy years in horticulture* (London, 1928, pp. 2-4), says

My first visit to Kew was in 1858, and impressions are difficult to describe. I was overwhelmed by its extent, and enthralled by the vastness of its collections. The summer was fine and warm (it was the year of Donati's comet), and the display of summer bedding beggars description.

Further comments on the intense competition among nurseries for novelties are in order. When the East Indian *Coleus blumei* and its varieties were introduced the market was brisk. Into this world Veitch and Sons pressed with vigor and the two brothers, Thomas and William Lobb, were among the most active of their collectors sending back novelties. Nor was the activity exclusively English. Belgium, in particular, but also France and Germany were pursuing untried horticultural subjects. It was inevitable that many of these horticultural discoveries were also new to the museum botanists busy describing the floras of the world. Theodore Hartweg went to Mexico in 1836, next visited Guatemala, the equatorial Andes, and re-

[1] K. Konig and J. Sims, *Ann. Bot.* 2:474-475. 1806, in translating the historical account of Joseph Dombey from the French of M. Deleuze.

turned to England in 1843. Though his search was for garden subjects his discoveries were described by George Bentham, master systematist, and it must have excited Lobb to read of Hartweg's Californian collections, as when Bentham reported *Ribes occidentalis*, "in collibus arenosis prope urbem Yerba Buena ad sinum San Francisco."[2]

William Lobb, then, was in the parade of collectors for great horticultural establishments lead by the heroic David Douglas who was followed by Thomas Coulter and Karl Theodore Hartweg of Karlsruhe. Contemporary with Lobb was the Scot, John Jeffrey, who was in northern California and Oregon in 1852-53, collecting especially tree seeds for the "Oregon Committee" of Edinburgh.[3] Jeffrey was in San Francisco during the winter of 1853, although two other botanical collectors, Andrew and William Murray, could not locate him.[4] A. F. Beardsley was then collecting for a Scottish nursery; he visited the Santa Lucia Range for seeds of the Bristle-cone Fir (*Abies bracteata*), following Hartweg's unsuccessful mission and Lobb's introduction of that species into England.[5] But we have no record that Lobb met Jeffrey, the Murray brothers, or Beardsley, so secretive were the movements of these plant hunters employed to search the mountains and valleys of California for plants in the days of gold.

It was not long after I moved into Room 3004 of the Life Sciences Building at Berkeley, to assist Professor Jepson in the preparation of manuscript for his Flora of California,[6] that William Lobb became a recurrent name. In an attempt to fix the type locality of Lobb's several discoveries I found the record tantalizingly sketchy. William Henry Brewer's single sentence said that Lobb was in California from 1850 to 1853 (*Botany of California* 2: 557, 1880), but several authors reported his death in San Francisco in 1863 (Veitch, 1906, *et al*), though Killip (1932, 3) gave his death year as 1857. On an early visit to San Francisco I sought Lobb's grave in Lone Mountain Cemetery and with the aid of the grave books found his headstone, broken in two, which recorded his death as May 3, 1864. Discrepancies came to light, too, in the printed accounts of his travels and collections. I collected notes from the

[2] George Bentham, *Plantae Hartwegianae*. Londini, Guil. Pamplin, 1849. p. 311.

[3] *Cf.* Coville, F. V., 1897, and J. T. Johnstone, *Notes from Royal Bot. Gard. Edinburgh* 20: 1-53. 1939.

[4] Andrew Murray (1812-78) is well known, *cf.* Britten and Boulger, 1931, though his American travels have been overlooked. He wrote from Bingham, Utah, May, 1873, on the basis of letter in Department of Botany, British Museum (Natural History). His brother William is much less known. According to E. L. Little, Jr., 1953, pp. 107-108, William Murray collected seeds of *Chamaecyparis lawsoniana* in California in 1854, but might this have been John Jeffrey? Or A. F. Beardsley?

[5] A. F. Beardsley "collected for Lawson & Co. of Edinburgh about 1852 to 1854, and probably for others later. He collected seeds rather than specimens" (W. H. Brewer, *Botany of California* 2: 557. 1880). The few letters in the Hooker Correspondence at Kew are most interesting. If a lode of business papers, letters, etc., could be traced, perhaps no other person of the decade would be so revealing for Beardsley "knew the right people". Yet evidently he was not highly successful.

[6] *Cf.* W. L. Jepson, *A flora of California* 2: 10. 1936, fn.

literature on Lobb's stations for endemic species whose restricted distributions might provide evidence for his itineraries, a search which progressed slowly since very few of Lobb's specimens now have data, and the older identifications were often unreliable. With my departure for Colorado this search for Lobb clues was abandoned, but with recent opportunities for visits to England I resumed the project. Many *lacunae* persist in the Lobb story but, in view of the enforced secretiveness of his employers and the evident destruction of his letters and the firm's records before their going out of business (Killip, 1932, 3), it is unlikely that large additions to the record will come to light. Perhaps some household in Cornwall where he was born and lived, and worked as a young man, may harbor some letters written from foreign lands, but Lobb may have been pledged to forego such letters. Only a single letter of his, written in San Francisco, has been located, and is reproduced here. No portrait is known and no surviving kinfolk have been traced. An untracked spoor is Lobb's connection, after severing his affiliation with Veitch and Sons, with Messrs. Low and Co. of Clapton Nursery about the year 1860. In particular we should like to know whether he reached Cedros Island where he is reported to have found *Delphinium cardinale,*[7] and we wonder where he crossed the coast ranges of southern California to reach such deserticolous species as *Chilopsis linearis* and *Phacelia campanularia,* and precisely how far north he penetrated into Oregon.

Lobb's part in the history of the Big Tree *(Sequoiadendron giganteum)* is both critical and veiled with some secrecy. Kellogg's account is certainly trustworthy, yet we should like to know the other side of the pursuit for those seeds and specimens that were whisked off to England for description by Professor John Lindley as *Wellingtonia gigantea!* Dear Dr. Parry, who was evidently with Lobb at Monterey just before Veitch's messenger set out for the Calaveras Grove, and therefore should have firsthand information, places almost criminal shades on Lobb's actions (1883). The history of the Big Tree remains to be written from original documents. Materials toward such a study are included here, but much remains in the contemporary newspaper record and in archives of Californiana. Farquhar (1948) has brought together important notes on the portraiture of the Big Tree. An album of the many illustrations of *Sequoiadendron,* both serene in its forest home and mutilated for display in New York and London, would make an interesting volume. Professor Emanuel Fritz, author of an exhaustive bibliography of the California Coast Redwood (1957), has in manuscript an extensive companion bibliography of the Big Tree which should be published without delay.

[7] *Cf.* Earle I. Wilde, "Studies of the genus Delphinium." *Cornell Univ. Agric. Exp. Sta. Bull.* 519: 49. 1931.

EARLY YEARS

William Lobb was born in 1809 at Perran-ar-worthal between Truro and Falmouth in east Cornwall, to John and Jane Lobb. The family lived at Lane End, a hamlet on the road to Bodmin near Pencarrow where his father, a carpenter, was employed at Pencarrow estate. Tradition remembers John Lobb more for his woodsman's talents than as a carpenter. His brother left the elder Lobb some property and in consequence his carpentry evidently suffered, and in due course he became a gamekeeper to Sir Charles Lemon at Carclew but died while William was still a lad. Soon we find William employed at Scorrier House where the owner, a Mr. Williams, encouraged the young gardener to seek employment with Veitch.[8] In 1837 he began working at the nursery of James Veitch Sr. at Killerton near Exeter, and on October 4th he exhibited cryptogams at the Cornwall Provincial Horticultural Society's show. J. C. Loudon reported the event in his *Gardener's Magazine* (13: 554. 1837): "For the best collection of cryptogamous plants, to Mr. William Lobb, Feock; these are beautifully dried, and apparently very perfect specimens." Lobb was found to be "quick of observation, ready in resources, and practical in their application" (Veitch, 1906, 37-40). He first went to Brazil in 1840 and sent back such attractive introductions as *Dipladenia splendens* and *Hindsia violacea*. The year 1842 brought the first notice of Lobb in the *Botanical magazine* when *Begonia coccinea* was illustrated, although without exact mention of his name. Beginning in 1843, however, and 89 times in all Lobb was to be cited in its pages (Hunkin, 1947, p. 35).

Returning to England in 1844 Lobb was reappointed by Veitch and again set out for Brazil in April. For the next four years he explored Brazil, Chile, Patagonia, Peru, Ecuador, and Colombia. Urban (1906) summarized his Brazilian travels and collections; Killip (1932), those of the northern Andes. During his visit home he assisted Francis P. Pascoe with his list of Cornish plants (cf. *Bot. Gaz. (London)* 1850), contributing two records.

Evidently Lobb did not forsake his interest in Cryptogams, for William Wilson of Warrington wrote W. J. Hooker, on June 21, 1849:

> I received the parcel yesterday afternoon and have already looked over Mr. W. Lobb's mosses — they do not contain many new species, and you will see what they are by the enclosed list. There are very few species from Patagonia — As yet I have not had time to compare them with authentic specimens, and it is possible that when I do so I may have to alter one or two of the names.[9]

Wilson enclosed a list of 45 mosses. One entry reads: "Leptostomum splachnoides (H & Arnt) fine spec[ime]n."

[8] J. H. Veitch (*Hortus Veitchii*, 1906) reports that William Lobb took employment with Mr. Stephen Davey of Redruth in 1837, but according to Hunkin (*J. Roy. Hort. Soc.* 72: 34. 1942) it was rather his brother Thomas who worked for Davey.

[9] Hooker Corresp., Royal Botanic Gardens, Kew [vol.] 40: [page] 370. ms.

Lobb arrived in San Francisco during the summer of 1849. What a discovery would be a letter of his written back to Cornwall! It is estimated there were about two thousand adults — few of them women — living in tents that summer. By June 1st about two hundred vessels were clogging San Francisco harbor detained or abandoned because their crews had fled to the mines. On October 27, 1849, Phipps Hornby, a relative of Capt. W. W. Hornby, R.N., of Knowsley, Liverpool, wrote W. J. Hooker from Valparaiso, Chile, as follows:

> It will be my Pleasure, as it is my Duty to afford every assistance in my Power to Mr. Lobb.
> I have a Ship at present, at San Francisco in California & shall most probably be enabled to keep a Cruizer pretty regularly about that Coast & the Island of Vancouver & the Captains shall have Instructions to give every aid in their Power to Mr. Lobb in the Prosecution of his Interesting Employment. I fear he has fallen on rough ground, but those he will meet with, in this Search after gold, will better envy his collections & I trust therefore neither interfere with or molest him. I am happy to say Mrs. Parker & my family are all well, we are preparing for a long voyage, first to Conception in Chile, south of this, & then for the Coast of Peru, Lima, Tahiti, & the Sandwich Islands. I wish Mr. Lobb were going with us to assist our Botanical Researches.[10]

The botanical collector Timothy Langdon Andrews arrived in San Francisco in November. Lobb may have met him there before Andrew's early departure for Monterey. The following spring Andrews accompanied Lobb on his trip into the Santa Lucia Range for Bristle-cone Fir seed. Veitch had instructed Lobb to collect seeds of several of the conifers discovered by Douglas and Coulter which were not yet introduced into English gardens. Dr. C. C. Parry, who had accompanied the Mexican Boundary Survey to San Diego, had travelled north to Monterey early in 1850. John Torrey, who had implemented Parry's appointment, wrote W. J. Hooker from Princeton, N.J., on June 20, 1850, relative to shipping the novelties that were being dispatched with the opening of the Great West, that "goods are very apt to be detained on the Isthmus, in consequence of the imperfect arrangements of forwarding that exist in the present rush to California."[11]

Veitch was pleased with Lobb's collections from the Santa Lucia excursion which must have reached as far south as the serpentine areas near Mission San Antonio de Padua where *Acanthomintha lanceolata*, an endemic labiate of those outcrops, was taken. Following the practice of Douglas, and of Coulter before him, Lobb evidently took advantage of the missions as places of lodging on his *jornadas* through the Coast Ranges. Parry accompanied Lobb on shorter trips in the vicinity of Monterey as on their visit to Mission San Juan de Bautista (Rodgers, 1942, p. 222). Certainly the high point of the year 1850 was the successful shipment of *Abies bracteata* to Veitch and its introduction into horticulture.

[10] Hooker Corresp. 70: 63.

[11] Hooker Corresp. 53: 395.

The year 1851 is wholly undocumented as to Lobb's travels except that he set out for northern California, again for conifers, and again we learn from the correspondence between Torrey and Hooker that there was a feverish interest in detecting possible new species. Torrey kept a close vigil on the all the exploring parties in the West. The United States Exploring Expedition under Charles Wilkes had sent a party to the headwaters of the Sacramento River, and Torrey wrote W. J. Hooker November 24, 1851, concerning a "Thujoid plant" that was to be described in April, 1953, as Incense Cedar (*Libocedrus decurrens* Torrey):

> I have been able to put up for you only two or three specimens, concerning which I desire your opinion. One is a Thujoid plant — Northern California. I first saw it in the collection of the U. S. Exploring Exped^{n} but without fruit. A few days ago I received some *bits* in fruit from a traveller lately returned from California. It seems to be a new genus, differing from Thuja but approaching Callitris. Have you anything like it from the N. W. Coast? You will also find specimens of the famous Red Wood or Taxodium? sempervirens which Endlicher makes the type of a new genus. I have long tried to get the fruit of this tree. Only two or three times have I been able to obtain the fertile aments, but in each case the seeds had fallen. A friend who long tarried in the region where the tree abounds says he never could find the seeds — nor did he see any seedling plants of the species. Do you think the genus a good one? I should be glad of your opinion of both these trees when you have leisure to examine their character.[12]

Here will be recognized the separation of the genus *Sequoia* from *Taxodium* by Endlicher. The correspondent's difficulty in finding seedlings of the Redwood seems strange indeed in view of the abundance of the seedlings today, but we must remember that he knew the coast forests unspoiled when the veterans stood in groves free from the second growth so familiar now.

The year 1852 found Lobb again in the mountains of northern California. He may have taken a coastwise vessel to the mouth of the Columbia, for Veitch (1906) particularly stresses his *return* from Oregon "through North California" collecting conifer seeds along his route. But we have no details of his Oregon collecting stations.

Lobb was in Monterey in the spring of 1853 and gave special attention to the *Ceanothi* that season, collecting at least five species in the vicinity which include three hybrids of great rarity known to this day only from the Monterey region (Van Rensselaer, 1942, p. 97). This was the year of the founding of the California Academy of Natural Sciences in San Francisco. Five doctors, including Dr. Albert Kellogg, a real estate man, and a school superintendant, met informally on April 4th to establish a "club" of like-minded men who hoped to share their enthusiasm for the natural sciences. The Big Tree loomed large in the attention of the Academy's members that year.

[12] Hooker Corresp. 53: 398.

LOBB AND THE BIG TREE

The full story of the first description of the Big Tree from specimens taken to England by Lobb will perhaps never be retrieved from the confused and incomplete records, but the principal threads of the story mesh as follows. In August 1833 Capt. Joseph Reddeford Walker and his party encountered "some trees of the red-wood species, incredibly large — some of which would measure from sixteen to eighteen fathom round the trunk and at the height of a man's head from the ground" (Zenas Leonard, *Narrative*, 1839). The second sighting, which has been disputed, was John Bidwell's in 1841, though he did not report it for many years. For that reason credit is usually given to the hunter, A. T. Dowd, who unquestionably saw the Big Trees in Calaveras Grove in the spring of 1852.[13] It was possibly this visit that furnished Dr. Albert Kellogg with the sterile specimens mentioned by Jepson (1946, p. 300) which Kellogg shared with Torrey and Gray. The specimen he sent to Torrey was evidently lost in the mails, and Asa Gray, puzzled by the cupressoid nature of the species, was reluctant to make a decision as to its taxonomic position among the Coniferae. Here Lobb enters the story. Let Kellogg speak:

> As historic truth demands, it is but just to say, I, myself, took Mr. Lobb to the California Academy of Sciences, and showed him the first specimens he ever saw of this marvelous, now world-renowned, Washington Cedar, which was so named by me, before he ever saw the tree. This fact is well known to the old charter members of the Academy, several of whom are still living (Kellogg/1882/, 19).

Lobb seized the day, and hastening we may say on wingéd feet, collected a good quantity of seed, herbarium specimens, and two living plants. In Lobb's account, quoted by Lindley, he tells of "a tree recently felled," from which he probably took his supply of seed. The tree "measured about *300 feet in length*, with a diameter, including bark, *29 feet 2 inches, at 5 feet from the ground* ..." Twenty one feet of the bark of this "vegetable monster" was taken to San Francisco for exhibition. Lobb immediately took passage on a ship bound for England. It is history that on Christmas eve, 1853, Professor John Lindley published the description of *Wellingtonia gigantea* in *Gardener's chronicle*. This was named for the "greatest of modern heroes" who "stands as high above his contemporaries as this Californian tree above all the surrounding foresters." The Duke had died the year before, and Lindley crowned a British hero, but unleashed American cross fire that was to consume hundreds of pages for decades to come.

[13] A. T. Dowd, "hunter" and employee of the Union Water Co., Murphys, Calaveras Co., gained a certain fame through the writings of J. M. Hutchings. For a good account of Dowd see anon. *Short account of the Big Trees of California*, p. 8, 1900, where his grizzly bear story is retold.

Thus, as Whitney (1869) remarked, "the honor and opportunity of naming [the Big Tree] was lost to American botanists." Even Englishmen were moved to write: "It is much to be regretted that any sinister motives should ever interfere with botanical science, so as to drive Botany from its high position, to occupy the low footstool of flattery" (Gordon, *Pinetum*, Suppl. p. 106). Less than a year after the founding of the California Academy of Natural Sciences the corresponding secretary read a letter from Dr. John L. LeConte, of the Academy's Philadelphia counterpart, which enclosed a copy of Lindley's description of *Wellingtonia* the previous December. The Minutes of the California Academy for March 27, 1854, read in part:

... the Society should maintain its ground. Months ago [Dr. Kellogg] proposed the name of Washingtonia, for the mammoth tree, but in waiting for further information respecting described species, we have lost it. After further remarks by the president and other members, it was unanimously — *Resolved*, That in view of the isolated condition of this Academy from other societies that we will regard every publication of new species which has been, or which may be made through the daily papers of this city, as substantial evidence of priority of discovery.

There were some facts concerning the sales of introductions by the Veitch nursery that bear on this contest. James Veitch wrote W. J. Hooker from Exeter on July 8, 1853, in part:

We have still several very fine interesting plants of William Lobb to bring out and I trust with my son in London we shall be better remunerated than we have been for some of our fine new plants already sent out the Profits on which have been more to others than to ourselves.[14]

Perhaps Hartweg, Jeffrey, and the lesser activities of the Murray brothers and Beardsley, were encroaching on Veitch's supremacy in the sale of exotics. In any event it is certainly understandable that a loyal employee would seize the opportunity to "scoop" a novelty, particularly when that novelty was the biggest tree yet seen by Man. Indeed, it is remarkable that the existence and description of such a "mammoth" was so long hidden. Tall tales of tall trees sprouted in many mining camps to be sure. Dowd's design of baiting his cronies to see the biggest grizzly ever seen so that he might lure them to see the Big Trees underscores the disbelief.

Asa Gray's perplexity adds to the little drama. On February 7, 1854, he wrote to W. J. Hooker:

Torrey being much interested in the Coniferae, I seldom look at them now, nor have I any Californian correspondent directly. Torrey has several. I sent him your letter, and I enclose the one I have just received from him. Some branches of Sequoia (Taxodium sempervirens) he sends have small and articulate leaves on very short stems. I am slow

[14] Hooker Corresp. 33: 429.

to believe that it could ever have such Juniper leaves as the *Wellingtonia* you speak of. Torrey should remember that you have (as I have also) the Sequoia growing in England, and could not confound it with anything else.[15]

To sum up, queried Gray, "But is not the *Wellingtonia* a *congener* of Sequoia?" Just before this Torrey wrote Gray — the letter relayed to Hooker with his own and dated New York, February 3, 1854:

That Wellingtonia of California (named, I suppose, by Lobb) is probably the same tree that Parry was lately speaking to me about. He saw specimens of it last week at the Acad. of Sc. Philad. It is said to grow to immense size even to 300 ft. in ht. & many ft. in diameter .... He brought me a small twig which looks like Juniper [half-inch fragment "This is a bit" inclosed with letter] ... Is it possible that the Wellingtonia can be a very old state of Sequoia?[16]

Asa Gray lamented to Hooker in a letter of March 28, 1854:

I have some pieces of the wood of the great *Wellingtonia* tree, which I estimate to be not older probably than the Christian era Torrey has no fruit, nor have I; but there are some cones in Philadelphia. The *wood* is very like that of the "Red-wood" (i.e. Taxodium sempervirens. I hope we shall get the male flowers, but I have no correspondents in California, and Dr. Torrey no very good or energetic ones.[17]

When the record is reviewed Jepson's *apologia* for Dr. Kellogg (Jepson, 1946, p. 300) pales a trifle. It was a period of intense and exciting discussion.

By the summer of 1854 several parties had set about to exhibit, P. T. Barnum fashion, the Big Tree in portable slab sections on both sides of the Atlantic. Farquhar (1948) tells this story from the printed record. The *entrepreneurs* approached the botanists. For example, Torrey wrote to Hooker on behalf of William H. Hanford, June 15, 1854, regarding exhibiting a section of bark, etc., at Sydenham, and adds, "The trunk as set up, is a grand,— I may say a sublime spectacle You can hardly conceive of the enormous size of the entire tree, till you stand at the base of the trunk ..."[18] A letter followed soon from William H. Hanford, delivered by C. A. Todd, to W. J. Hooker, June 21, 1954:

I now take the liberty of writing you in relation to the "Gigantea Wellingtonia" or "Mammoth Tree" from California. The bark of the tree was carefully taken from the trunk while standing commencing at eight feet from the base and stripping it to the height of about forty five feet, this gives us five sections or zones ...

I presume it will be unnecessary for me to go into minute description as Mr. Lobb (the Collector for Mess. Veitch of the Exotic Nursery Chelsea and of Exeter) has seen the tree while growing in its native forest and has also seen the bark (or three sections of it) while set up for exhibition in San Francisco and can doubtless describe it more particularly than I should be able to do with my pen.

[15] Hooker Corresp. 64: 179.

[16] Hooker Corresp. 53: 400.

[17] Hooker Corresp. 64: 181.

[18] Hooker Corresp. 53: 401.

My object now is either to bring this great natural curiosity to England for the purpose of exhibition or to sell it to the Director of Sydenham Palace . . . . should they not wish to purchase will you inform me if there is any building in or near London which could be obtained . . . for the exhibition of it and what success you think such an exhibition would be likely to meet with there.[19]

Whitney is right when he said that "no other plant ever attracted so much attention or attained such a celebrity within so short a period" (1869, p. 141). By this date the references to the Big Tree in scientific works and journals "number between one and two hundred ... innumberable articles in popular periodicals and books of travel, in various languages; probably there is hardly a newspaper in Christendom that has not published some item on the subject."

LATER YEARS

Lobb's return to California in 1854 must have been in the early spring since his collections for Veitch were spring flowering chaparral species with a few that came from the margin of the Colorado Desert, perhaps about the Whitewater "bench" or Morongo Pass where, for example, *Chilopsis linearis* comes nearest to the coastal plain. Did Lobb disembark at San Diego or San Pedro? The time lapse between his departure from England and the flowering dates of the plants he collected in southern California suggest that he did. If he disembarked at San Diego he may have taken the opportunity of visiting Cedros Island, then an object of interest,[20] but we are not at all sure that this report is accurate. Nor have we any record of his activities in or around San Francisco at this time. He may have been, indeed, *persona non grata* in that city. In any event discussion of the generic status of the Big Tree was brisk at the Academy that year. Dr. W. P. Gibbons exhibited specimens of the Big Tree and the Redwood and emphasized their congeneric status. That was at the Academy meeting of February 12; on April 9th, Dr. Kellogg exhibited drawings he had made of the staminate cones of the Big Tree; and at the May meeting he again urged the congeneric nature of the two trees. When the Fall meetings resumed H. G. Bloomer donated a cone of "Taxodium giganteum" to the

[19] Hooker Corresp. 64: 275.

[20] Although first discovered by Francisco Ulloa in 1539 and named *Isla de Cedros* for *Juniperus californica* found there, and again visited in 1846 by H.M.S. *Herald,* Cedros Island was not scientifically ransacked until the extended stay of Dr. John Allen Veatch (1808-1870). Veatch collected plants and sent them to Dr. Kellogg who described them. Veatch himself published a popular account of his trip (*Hesperian* (San Francisco) 3: 529-534. 1860). The date of Veatch's visit is variously given as 1854 (P. A. Munz, *Leafl. W. Bot.* 7: 70. 1953), "about 1858" (S. W. Geiser, mimeo. letter, 18 Sept. 1942), and 1859 (E. W. Nelson, *Mem. Nat. Acad. Sci.* 16: 87, 141, 149. 1921), but in any event Lobb evidently had no connection with Veatch's expedition. Cf. Geiser in *Field & Lab.* 27: 228-229. 1959.

Academy's cabinet. There is no record of Lobb's having attended the Academy's meetings, but not all the names of visitors are recorded in the Minutes. For example, on January 16, 1854, the Minutes read: "visitors, A. M. Jackson, corresponding members and others."

Nothing is known of Lobb's whereabouts during 1856. He may have been ill, perhaps from some lingering infection contracted in the tropics. A letter from James Veitch, Nursery, Exeter, January 2, 1857, to W. J. Hooker, reads in part:

> You will be pleased to hear we have also heard from *William Lobb* from California advising us of his having sent off 3 boxes but whether they will arrive by the Mail now due or the next I cannot say. One consignment of his sent off 7 or 8 months since or more round Cape Horn has never reached us & I conclude it is lost. it was mostly seeds of Pinus Lambertiana . . . W. Lobb writes merely a few lines saying he will write again by next mail we hear he has been ill & his writing appears *shaky* and I am inclined to think it probable he will now return ...[21]

and Veitch remarks on the collecting done by A. F. Beardsley, adding, "Lobb fared just as Beardsley in his first attempt to get the Picea Bracteata. We have now a nice stock of it, plants 1 foot high. I think we sent it to Kew." Veitch's letter was followed by another, of January 12, 1857, to Hooker, commenting:

> We have at last heard *satisfactorily* from *Wm* Lobb with a consignment not much novelty but a few good things no dried specimens & he does appear to have collected *any* tho' we told both him and Thos Lobb that when they could do so they were to collect any novelties for you. W. Lobb's letter to me the only *one* I have received for a long time merely gave a list of what he had sent without any particulars of his Journey or discoveries or his future movements but we expect to hear again by the next mail—we are also expecting shortly to hear from Thomas Lobb from Borneo.[22]

Meanwhile Thomas Bridges had arrived in California. Bridges, like Lobb, had collected previously in Brazil and had contested for novelties though he lost recognition for this introduction of the *Victoria amazonica*. He had come to California in November, 1857, explored the Sierras, settled in San Francisco, and finally after a recurrent illness died at sea and was buried in San Francisco.[23] During 1857 both Bridges and Beardsley were collecting in the Sierra Nevada and the Santa Lucia Range. Lobb terminated his seventeen years with Veitch this year. On May 19, 1857, A. F. Beardsley wrote to W. J. Hooker from San Francisco that he could not collect both "Picea

[21] Hooker Corresp. 37: 587.

[22] Hooker Corresp. 37: 588.

[23] The career of Thomas Bridges (1807-1865) was typical of the professional collector of the last century, and we hunger for more letters written in the field. Jepson (*Madrono* 2: 84-88. 1933) published a long letter to W. J. Hooker, dated San Francisco, May 5, 1858. To the biographical references given by Jepson may be added I. M. Johnston's paper in *Contr. Gray Herb.* 81: 98-106. 1928.

nobilis" and "picea bracteata" since they were six hundred miles apart, and asked Hooker which he would like, adding:

> Mr. Bridges and Mr. Wm. Lobb, collectors, are both in the country, and will probably throw their collections into the market. It therefore becomes me (without correspondents) to be cautious how I go to too great an outlay.
>
> You say that nurserymen are jealous of others obtaining the same things. The best assurance I could give them would be to send all my collections to one man; who would solicit them, and sudjest[*sic*]the best varieties to be sent and the quantities desirable of each. Much valuable time might be saved, and perhaps the success of the expedition might depend upon seeking only two or three things. The distances and mode of traveling in this country seems to be little understood in Europe ...[24]

James Veitch refers to a "shrubby" *Penstemon* in a letter to Hooker from Exeter, dated August 27, 1857, explaining that it grows two and a half feet high, without doubt *Penstemon antirrhinoides,* distributed as "P. lobbii." Veitch wrote:

> I think I have dried specimens from W. Lobb. I will endeavour to find them & send you one tomorrow—I also send specimen of an Oenothera. W. Lobb 411, from Plains of San Gabriel South California. This I have sent you before, and I think I have the name from you in one of our letters ... as we are ready to send out the seed this coming season we must give it a name.[25]

The only letter written by Lobb so far traced was addressed to W. J. Hooker, from San Francisco, February 4, 1858:

> I write you a few lines by this route to inform you that I have forwarded to Aspinwall to be shipped on the West India Steamer Five cases Seeds and Specimens of plants marked and numbered L Nos 1,2,3,4,5, with full directions. Also a letter by the same conveyance with full particulars respecting the collection.
>
> Should this reach you before the arrival of the Steamer at Southhampton due March 15th probably it would be advisable that some person should see the cases opened at the customhouse and the contents carefully handled.
>
> I am / Sir / your obdnt Servant / W. Lobb[26]

There is no trace of the second letter referred to. Beginning in the year 1858 Lobb's name appears in the San Francisco directories, first as a resident at the se. corner of California and Montgomery streets, then from 1860 his address is given as "28 U. S. Court Building."[27] From George Gordon there is evidence that he collected for Messrs. Low and Co. of Clapton

[24] Hooker Corresp. 64: 19.

[25] Hooker Corresp. 37: 595.

[26] Hooper Corresp. 64: 267.

[27] The Society of California Pioneers, San Francisco, possesses five San Francisco directories for the period 1858-63. Five pictures of the Big Tree are also in its collections (*Annual Publication,* 1955, p. 22) by the artists T. A. Ayres, E. Vischer, and H. J. Breuer.

Nursery. In the Supplement to the *Pinetum* (1862) Lobb is credited with "Abies pattoni," "Picea grandis," and "Picea Lowiana." Of particular interest is the record of Lobb's sending the Torrey Pine, the rarest of California pines, for the Clapton Nursery in 1860. Perhaps this seed was gathered on what may well have been his last collecting trip: to the Soledad River north of San Diego. Again Professor Lindley described Lobb's collection as a new species, *Pinus lophosperma,* but in this action he had been preceded by Parry five years before. There is no evidence that Lobb collected other localized species in southern California on this trip.

By 1862 there was the ferment of the State Survey with William Henry Brewer travelling "up and down California" observing, recording, collecting toward what was to be the *Botany of California,* written in collaboration with Asa Gray and Sereno Watson. As a member of the State Survey, Brewer met Beardsley in October, 1861, in a tavern in Napa Valley whence Brewer had repaired "to read the news."[28] But Brewer found a few cronies in his travels, and wrote to Hooker from San Francisco on August 12, 1862:

> In answer to your inquiry 'if there are good collectors or accomplished Botanists, *resident* in this state'—I can only say unfortunately *No.* Dr. A. Kellogg has described a number of new species, but is resident in this city and does not collect to any extent. H. G. Bloomer Esq. is an Amateur much interested in the Botany of the state, but also resident here and no collector—There are a number of men who make small collections but no systematic collectors, save connected with our Survey—I am collecting as fully as possible & hope some time to have some good duplicates—A young man has recently arrived, a Mr. Bolander, now of this city who I think will collect *considerable.*[29]

Thus in Brewer's roster of California botanists Lobb's name does not appear although it was listed that year as "Lobb, William, botanist ..." in the San Francisco directory. He was no longer active. The San Francisco *Daily Alta Californian* for Thursday, May 5, 1864, carries the only record of Lobb's later years:

> In this city, May 3rd, William Lobb, a native of England, aged 59 years.
> Friends and acquaintances are respectfully invited to attend the funeral this day, at 11 o'clock A.M. from St. Mary's Hospital, without further notice.

According to Veitch (1906, p. 40) his illness was paralysis; he was stricken in 1863 and lost the use of his limbs. Bridges was in San Francisco at the time of his illness and death, but we have found no mention of Lobb in any personal vein in his letters. Lobb was buried in the Public Plot at Laurel Hill Cemetery (renamed from Lone Mountain Cemetery in 1862), and there he rested until Alice Eastwood enlisted the aid of the California Botanical

[28] W. H. Brewer, *Up and Down California in 1860-1864,* edited by F. P. Farquhar (Yale Univ. Press, 1930), pp. 218-219.

[29] Hooker Corresp. 65: 7.

Club in moving his grave to the South Ridge Lawn. There in 1934 I photographed his headstone, then broken in the middle, which reads:

In memory of William Lobb.

a native of Cornwall, England

who died at San Francisco

May 3, 1864 aged 55 years.

The record of his age as 59 years, published in the newspaper account, is of course incorrect. In 1942 at Devoran Parish Church, near Truro, England, a memorial tablet was erected:

> In memory of Thomas Lobb buried in this churchyard 3rd May 1894 and of his of elder brother William, who died in America: two collectors of plants from foreign countries who rendered distinguished service to British horticulture.[30]

At the same time four shrubs were planted in the churchyard.

WILLIAM LOBB'S POSITION IN CALIFORNIA BOTANY AND HORTICULTURE

As a botanical explorer Lobb was the first to collect at higher elevations in the Sierra Nevada where he found *Eriogonum lobbii, Nama lobbii,* and other species described by Torrey and Gray. Professor Jepson noted in his Field Book (Vol. 35, p. 115, ms.): "*Eriogonum Lobbii* T. & G. is a fine species to have named for one and will well memorialize Lobb's expedition into the Sierra Nevada and show that he attained a considerable altitude." Had he carried his exploration of the boreal flora of the Sierra Nevada further he would certainly have anticipated Brewer, Bolander, Bridges, Muir, and others who were to dislodge new species from those summits. Elsewhere in California Lobb followed the well-worn trails for the most part, all previously botanized. His trips into the inner South Coast Ranges of Monterey County followed the way of Douglas and Coulter along the Camino Real. But Lobb's detection of the very rare natural hybrids in *Ceanothus* which have not since been found outside the Monterey region where he botanized in 1853 attests to his lynx-eyed collecting.

Overshadowing all stands Lobb's role as messenger of the Big Tree, his taking specimens to one of Britain's leading botanists who immediately

[30] *Cf. Childrens Newspaper,* 31 Oct. 1942, p. 2. and *Jour. Royal Hort. Soc.* 72: 35. 1947.

saw its uniqueness and described it, naming it albeit militantly, in the leading horticultural journal of the day. One wonders what might have been its history had Lindley named it "Francisdrakea," for example, instead of *Wellingtonia.* That an American botanist, Torrey or Gray, if not Kellogg, would soon have botanically described the Big Tree had not Lindley preëmpted the role of author, does not diminish Lobb's accomplishment.

From the horticultural standpoint Lobb, first of Veitch's world travelers, twenty two in all, won his laurels early with his tropical American greenhouse introductions, then in California with the conifers, flowering shrubs, and beautiful herbaceous species, such as the Cardinal Larkspur. Although Lobb was laid to rest in potter's field, quickly forgotten, and a century later remained unnoticed in a recent history of California gardens,[31] his living contribution to English gardens, either as horticultural subjects or taxa new to botanical science, numbers fifty eight California species. On the edge of a cricket field at Oxford stands a circle of "Wellingtonias" said by tradition to have been planted by William Hart Baxter. The Caucasian pigeons hoot in the old horse-chestnuts hard by, chinese weeping willows line the banks of the Cherwell down the hill, and relaxed birches wait in contemplative peace. On many an estate in England, in an old established garden near Philadelphia, and on the shores of Lake Geneva, stand towering *Sequoiadendron* whose seed Lobb first delivered to horticulture.

[31] Victoria Padilla, *Southern California gardens. An illustrated history.* (Berkeley, Univ. Calif. Press, 1961).

CHRONOLOGY

1809

Born in Perran-ar-worthal, East Cornwall, the son of John and Jane Lobb; baptized March 12.

1837

Went to James Veitch's nursery at Killerton, near Exeter, at the suggestion of Mr. Williams of Scarrier House, whose gardener Lobb had been

Oct. 4. Exhibited Cryptogams at Provincial Horticultural Society, Cornwall.

1840

Made first trip for Veitch: to Brazil.

1842

Botanical magazine, t. 3934, first refers to William Lobb, though not by name.

1843

Lobb's name first appears in *Botanical magazine* t. 3990.

1844

Returned to England.

1848

Aided F. P. Pascoe with a history of Cornish plants: contributed two new records.

1849

Summer. Arrived in San Francisco.

June 21. Letter from W. Wilson, Warrington, to W. J. Hooker, regarding Lobb's Patagonia mosses.

Oct. 27. Letter from Phipps Hornby, Valparaiso, Chile, to W. J. Hooker, assuring him he will "afford every assistance in my Power to Mr. Lobb."

November. T. L. Andrews arrived in San Francisco, but soon moved to Monterey.

1850

Lobb visited Monterey, botanizing in the vicinity on short trips, some with C. C. Parry, for example, about Mission San Juan Bautista.

Accompanied in the Santa Lucia Mountains by T. L. Andrews.

Visited Inner South Coast Ranges, at least as far as the serpentine outcrop in the vicinity of Mission San Antonio de Padua.

1851

Collected in northern California, getting *Pinus monticola,* etc.

1852

Collected in Oregon and on Columbia River.

June. Kellogg received specimens of Big Tree before this date, *teste* W. L. Jepson.

1853

In Monterey prior to a trip to the Sierra Nevada for Big Tree in Calaveras Grove.

John Jeffrey in San Francisco—did he meet Lobb?

July 8. Letter from James Veitch, Exeter, to W. J. Hooker, mentioning "several very fine interesting plants of William Lobb to bring out."

Lobb returned to England.

Dec. 24. *Gardener's chronicle* published *Wellingtonia gigantea* Lindley.

1854

Visited "Los Angeles plain," Mission San Gabriel.

Collected some deserticolous species, perhaps in Morongo Valley, Riverside County.

Feb. 3. Letter from John Torrey, New York, to W. J. Hooker concerning "Wellingtonia of California (named, I suppose by Lobb)."

Feb. 7. Letter from A. Gray, Cambridge, to W. J. Hooker, asking if "Wellingtonia" is not "a congener" of *Sequoia sempervirens.*

Feb. 11. *Ill. London News* featured the "Wellingtonia" discovery quoting from *Gardener's chronicle* of Dec. 24, 1853.

March 27. Members of California Academy of Sciences meeting disagreed with the Assistant Secretary of Smithsonian Institution on the priority of Lindley's name "Wellingtonia."

March 28. Asa Gray, Cambridge, wrote to W. J. Hooker reporting that he has word of "Wellingtonia," but has no correspondent in California to send specimens of cones, etc

June 5. Letter from John Torrey, New York, to W. J. Hooker regarding *Sequoia gigantea.*

June 21. Letter from Wm. H. Hanford, New York, to W. J. Hooker about the stripped-bark tree exhibited in New York City.

June 28. Descaisne communicated Big Tree specimens to Societe Botanique de France, received from French consular agent at San Francisco, and published (July) on the congeneric status of Big Tree and Redwood.

Sept. Asa Gray agreed to the congeneric status of the two in Silliman's *American journal of science.*

1855

William Lobb reportedly visited Cedros Island.

Feb. 12. Dr. W. P. Gibbons exhibited specimens of Redwood and Big Tree emphasizing their congeneric nature.

April 9. Kellogg exhibited his drawings of "male flowers," etc., of Big Tree.

May 7. Kellogg reported on the generic status of Big Tree.

Nov. 19. Minutes of the Academy record: "From Mr. H. G. Bloomer, a cone of Taxodium giganteum."

Dec. T. A. Ayres made his crayon drawing of "The Mammoth Tree Grove. The Hotel."

1856

November. Thomas Bridges, age 49, arrived in California.

1857

Jan. 2. Letter from James Veitch, Exeter, to W. J. Hooker, that Lobb has written "merely a few lines," and that evidently "he has been ill."

Jan. 12. Letter from James Veitch, Exeter, to W. J. Hooker that Lobb wrote "satisfactorily," and "with a consignment."

May 19. Letter from A. F. Beardsley, San Francisco, to W. J. Hooker that both Thomas Bridges and William Lobb are in the country "and will probably throw their collections into the market."

May 21. Letter from James Veitch, Exeter, to W. J. Hooker mentions a *Viola* sent by Lobb.

Aug. 18. Letter from James Veitch, Exeter, to W. J. Hooker mentions a hardy *Penstemon* sent by Lobb.

Aug. 27. Letter from James Veitch, Exeter, to W. J. Hooker mentions *Penstemon* and *Oenothera* from Lobb.

In the Santa Lucia Mts. (?) again for *Abies bracteata.*

William Lobb terminated his seventeen year engagement with Veitches.

1858

"W. Lobb, naturalist, dw[el]l[s] SE cor[ner] California and Montgomery [streets] (San Francisco Directory for 1858, p. 185)."

Feb. 4. Letter from William Lobb, San Francisco, to W. J. Hooker stating that he is sending five cases of seeds and specimens on the West India Steamer via Aspinwall.

Spring. Thomas Bridges returned to California from British Columbia and settled in San Francisco.

May 5. Letter from Thomas Bridges to W. J. Hooker telling of his botanizing in Mariposa County, etc.

1859

Feb. 14. Letter from Joseph Paxton, Sydenham, to W. J. Hooker communicating promised seed of "Wellingtonia gigantea."

1860

"Lobb, William, botanist, 28 US Ct Building."

Nov. 12. Letter from Richard Owen, London, to W. J. Hooker thanking him for the "beautiful young Wellingtonia."

1861

"Lobb, William, botanist, 28 US Ct Building."

April 30. Letter from Richard Owen, Sheen Gate, to W. J. Hooker, reporting that his seedling Wellingtonia "battled the hard winter bravely."

1862

"Lobb, William, botanist, dwl 28 US Court Bdg."

Aug. 12. Letter from W. H. Brewer, San Francisco, to W. J. Hooker mentions that H. N. Bolander is the only active botanist then in the state.

1863

"Lobb, William, botanist, dwl 28 Federal Building," San Francisco Directory for the year commencing Oct. 1863.

1864

May 3. William Lobb died of paralysis.

May 5. Buried in the Public Plot, Laurel Hill Cemetery, after funeral services at St. Mary's Hospital, eleven o'clock a.m. Thomas Bridges in San Francisco at this time.

Oct. 20. Letter from W. H. Brewer, San Francisco, to W. J. Hooker, regarding the Big Trees of the southern Sierra Nevada and stating that they are "abundant" locally.

1927

Dec. 1. William Lobb's grave removed from "Public plot" to "South Ridge Lawn" through the efforts of Miss Alice Eastwood.

1939–1940

Remains removed to crypt at Cypress Lawn Cemetery.

1942

Memorial tablet to William and Thomas Lobb placed in Devoran Parish Church near Truro, England.

LIST OF HORTICULTURAL INTRODUCTIONS AND NEW SPECIES

The species are listed in taxonomic sequence either by the name under which they were introduced into horticulture or as proposed first as new species. Synonyms are given only when they assist in following the concept through the literature. *Bot(anical) Mag(azine)* references are cited by t(abula) or plate number. Other references will be readily identified from their abbreviations. Finally, modern botanical names are printed in small capitals.

PINACEAE

Abies amabilis sensu Veitch. Collected in "North California" in 1852, Veitch, *Manual* 244. 1900. = ABIES MAGNIFICA Murr.

Abies bracteata D. Don. "Dr. Coulter first discovered it in the Andes of St. Lucia [Santa Lucia Mts., Monterey Co.] ... Douglas met with it at 6000 feet on the California mountains, in latitude 36° north. ... It was left for Mr. William Lobb, Messrs. Veitch's indefatigable American collector [to bring home perfect seeds]" *Bot. Mag.* t. 4740. 1853. cf. *Gard. Chron.* July 9, 1853, p. 435. Veitch (*Manual*, p. 244, 1900) called this fir "one of the first fruits of the expedition." For the basis of accepting *A. bracteata* as the valid name for this fir see W. A. Dayton, *Rhodora* 54: 74-76. 1952. = ABIES BRACTEATA D. Don.

Abies grandis (Dougl.) Lindl. Seed collected in "North California" in 1852, according to Veitch, *Manual*, p. 244. 1900. = ABIES GRANDIS (Dougl.) Lindl.

Abies nobilis (Dougl.) Lindl. Seeds collected on trip to Columbia River in 1852, according to Veitch, *Manual*, p. 244. 1900. = ABIES PROCERA Rehd.

Abies pattoni Jeffrey "Mr. William Lobb found it in abundance, on the highest peaks of the Sierra Nevada, near the head waters of the north tributary of Feather River, and more to the south, towards Lake Bigler; forming immense trees, in habit of growth and general appearance much resembling the 'Deodar', but being more thickly branched, and densely clothed with foliage; by far the handsomest tree." Gordon, *Pinetum, Suppl.* p. 7. 1862. = TSUGA MERTENSIANA (Bong.) Carr.

Picea grandis Loudon "This kind must not be confounded with a new and very distinct species sent to Messrs. Low and Co. of the Clapton Nursery, by their collector, Mr. William Lobb, from California, under the misapplied name of *Picea grandis*, and which kind I propose calling *Picea Lowiana* [*Abies concolor* var. *lowiana* (Gordon) Lemm.], in order to distinguish it from the true *Picea grandis* of Douglas [*Abies concolor* (Gordon) Hoopes], as much confusion has arisen both in the Nurseries and in private collections from the misapplication of the name (*grandis*) by Mr. Lobb. The original plant of *Picea grandis*, raised from Douglas's seed in 1831, and growing at Elvaston Castle, places the identity of the true *grandis* beyond any doubt." Gordon, *Pinetum, Suppl.* p.53, 1862 = ABIES CONCOLOR (Gordon) Hoopes.

Picea Lowiana Gordon "This new and very distinct species appears to have been first discovered by Mr. William Lobb ... It is to be found in nearly all the Nurseries, and in many private collections, under the misapplied name *grandis*, owing to Mr. Lobb having transmitted a large quantity of its seeds to Messrs. Low and Co. of the Clapton Nursery, under the wrong name." Gordon, *Pinetum, Suppl.* p. 55, 1962. = ABIES CONCOLOR var LOWIANA (Gordon) Lemm.

Pinus lophosperma Lindl. "A magnificent tree, somewhat resembling *Pinus Sabiniana*, with very glaucous, stout shoots, found in Lower California by Mr. William Lobb, who transmitted seeds of it to Mr. Hugh Low, of the Clapton Nursery, in the early part of the year 1860." Gordon, *Pinetum, Suppl.* p. 69. 1862. Important evidence that Lobb visited San Diego County where this pine of very restricted range grows. = PINUS TORREYANA Parry.

Thuja lobbii Hort. and *T. lobbiana* Hort. Seeds first received at Exeter in 1853 and distributed under the above names. "In 1852 he made an excursion to the Columbia river and Oregon where he succeeded in obtaining seeds ... [of] the beautiful *Thuia* provisionally named after him" Veitch, *Manual*, p. 244. 1900. = THUJA PLICATA Donn.

TAXODIACEAE

Wellingtonia gigantea Lindl. "First introduced to Europe through William Lobb, who sent seed and a living specimen to Exeter in 1853" *Hortus Veitchii*, p. 346. 1906. *Cf.* Gordon, *Pinetum, Suppl.* p. 106, 1862, but no mention of Lobb, nor in ed. 2., p. 415, 1875, where David Douglas singularly is credited with its discovery! According to Veitch, *Manual*, p. 277. 1900, Lobb returned with a "large quantity of cones and seeds" and "two living plants and herbarium specimens" in the autumn of 1853. = SEQUOIADENDRON GIGANTEUM (Lindl.) Buchholz.

CUPRESSACEAE

Juniperus pyriformis Lindl. "Introduced in 1853, at the same time as *Sequoia gigantea* ... discovered on the mountains of San Barnardino[sic] in California, ... and distributed under Dr. Lindley's name, Juniperus pyriformis" *Hortus Veitchii*, p. 340. 1906. Earlier the source of the Lobb seeds was given as "North California" on route from Columbia River in 1852, in Veitch, *Manual*, p. 244. 1900. = JUNIPERUS CALIFORNICA Carr.

TAXACEAE

Torreya myristica Hook. "A native of the elevated regions in the Sierra Nevada, California, by William Lobb rediscovered, and seeds and specimens sent in 1851" *Hortus Veitchii*, p. 347. 1906. *Bot. Mag.* t. 4780. 1854, and Gordon, *Pinetum*, p. 327. 1858. = TORREYA CALIFORNICA Torr.

LILIACEAE

Calochortus elegans var. *lobbii* Baker in *J. Linn. Soc. Bot.* 14: 305. 1874. "Oregon" *Lobb* 257. = CALOCHORTUS TOLMIEI Hook. & Arn.

Calochortus Maweanus Leichtlin ex Baker, *J. Linn. Soc. Bot.* 14: 305. 1874, as to "California" 1857, *Lobb* 242 at Kew. = CALOCHORTUS CAERULEUS (Kell.) Wats.

Calochortus lobbii Purdy, *Proc. Calif. Acad. Sci.* ser. 3. *Bot.* 2: 122. 1901. Type (K) is *Purdy s. n.*, from Mt. Jefferson, Marion Co., Ore., Aug. 1897. Though Purdy said "so far found only on Mt. Jefferson, Oregon," he did not provide evidence for Lobb's visiting that peak. The name as published by Purdy is a transfer of *C. elegans* var. *lobbii* Baker but Ownbey, *Ann. Missouri Bot. Gard.* 27: 420. 1940, concluded that Purdy's plant represented a distinct species. = CALOCHORTUS LOBBII Purdy.

Calochortus minimus Ownbey, *Ann. Missouri Bot. Gard.* 27: 430. 1940. Last coll. cited is "California," 1857, *Lobb* 243 (Kew) under "county uncertain" but very likely taken in vicinity of Calaveras Grove in 1853 where this mariposa has subsequently been collected. = CALOCHORTUS MINIMUS Ownbey.

Iridaceae

Iris douglasiana Herb. Lee W. Lenz (*El Aliso* 2: 148. 1950) suggests with caution that Lobb may have been responsible for the seed supplied to Veitch but remarks that this Iris is not mentioned among the introductions listed in *Hortus Veitchii*. In view of the fact that Lobb terminated his collecting for Veitch in 1857 and sent seeds to Messrs Low and Co. of Clapton Nursery after that or as late as 1860 it seems highly improbable that a Veitch introduction of 1873 has anything to do with Lobb.

Fagaceae

Castanopsis chrysophylla (Hook.) A. DC. "Seed sent from California by William Lobb, probably about the year 1853" *Hortus Veitchii* p. 393. 1906. = CASTANOPSIS CHRYSOPHYLLA (Dougl.) A. DC.

Polygonaceae

Centrostegia leptoceras Gray "plains of San Gabriel [Los Angeles Co.], Lobb in herb. Kew." = CHORIZANTHE LEPTOCERAS (Gray) Wats.

Eriogonum lobbii Torr. & Gray "California, Lobb in herb. Hook. no. 190" at Kew is the type. Eastwood (*Leafl. W. Bot.* 5: 155. 1949) suggests type was taken on Donner or Carson passes, over which there were roads to the Nevada mines. = ERIOGONUM LOBBII Torr. & Gray.

Portulacaceae

Spraguea umbellata Torr. "It has been introduced alive to the gardens of Messrs. Veitch ... through Mr. William Lobb" *Bot. Mag.* t. 5143. 1859. *Cf. Hortus Veitchii* p. 437. 1906. = SPRAGUEA UMBELLATA Torr.

Ranunculaceae

Delphinium cardinale Hook. *Bot. Mag.* t. 4887. 1855. Hooker had both garden grown plants from Lobb's seed and his herbarium collection (cf. Ewan in *Univ. Colorado Stud.* ser. D. 2: 190–192. 1945, where the problem of tracing Lobb's seed collection is taken up in some detail). *Hortus Veitchii*, p. 419. 1906. = DELPHINIUM CARDINALE Hook.

Ranunculus hydrocharis f. *lobbii* Hiern. *J. Bot.* 9: 66. 1871. "Oregon, W. Lobb! 1852, n. 249" is the type at Kew. W. R. Drew (*Rhodora* 38: 16. 1936) listed the names and references. = RANUNCULUS LOBBII (Hiern) Gray.

PAPAVERACEAE

Dendromecon rigidum Benth. "raised from seed sent home by William Lobb" *Hortus Veitchii* p. 363. 1906. *Cf. Bot. Mag.* t. 5134. 1859. = DENDROMECON RIGIDUM Benth.

Eschscholtzia tenuifolia Hook. *Bot. Mag.* t. 4812. 1854, based on garden grown plants from seed collected by Lobb and made available to Hooker by Veitch. Not *E. tenuifolia* Benth. = ESCHSCHOLTZIA LOBBII Greene.

FUMARIACEAE

Dicentra chrysantha (Hook. & Arn.) Walp. "originally discovered by David Douglas, and subsequently by William Lobb" *Bot. Mag.* t. 7954. 1904. "flowered in 1852 from seed sent from California by Lobb" *Hortus Veitchii*, p. 419. 1906. = DICENTRA CHRYSANTHA (Hook. & Arn.) Walp.

SAXIFRAGACEAE

Ribes lobbii Gray Some disagreement exists in the literature: "A native of California, where it was first detected by naturalists of ... Beechey's Surveying Voyage; and it has since been sent by Mr. Lobb." *Bot. Mag.* t. 4931. 1856. J. T. Johnstone (*Notes Roy. Bot. Gard. Edinburgh* 20: 12. 1939) said John Jeffrey discovered this species about Fort Victoria in April, 1852, but W. H. Brewer and S. Watson (*Botany of California* 1: 205. 1876) suggested Lobb discovered the *Ribes* "probably from the northern part [of California] near the coast." Finally, T. S. Brandegee (*Zoe* 4: 152. 1893) wrote "shady canons, Pacific Valley and Sur River, *Eastwood.* Fruit very large. As Lobb is known to have collected in the Santa Lucia Mountains, this probably the locality of the type." = RIBES LOBBII Gray.

ROSACEAE

Chamaebatia foliolosa Benth. "native of the 'higher parts of the Sierra Nevada, as well as the sides of the foothills. ... and the mountains of Sacramento, in California, and was first discovered by Colonel Fremont, in 1844,' ... and Messrs. Veitch and sons ... have the credit of importing live plants, sent by their collector ..." *Bot. Mag.* t. 5171. 1860. Though not named the Veitch collector was surely Lobb who may well have brought living plants from the Calaveras Grove in 1853 when he returned in haste to England with the *Wellingtonia* cones = CHAMAEBATIA FOLIOLOSA Benth.

LEGUMINOSAE

Lupinus aridus var. *lobbii* Wats. *Proc. Amer. Acad.* 8: 533. 1873, based on *Lobb* 264, without locality, but perhaps from Donner or Carson passes, collected in 1857, according to Eastwood, *Leafl. W. Bot.* 5: 155–156. 1949. = LUPINUS LOBBII Gray ex Greene. *Proc. California Acad. Sci.* 2: 299. t. 73. 1863.

Lupinus cervinus Kell. based on *Lobb* 119 from Santa Lucia Mts. collected in 1850 when he botanized there with T. L. Andrews who may indeed have communicated this specimen to Albert Kellogg. M. K. Curran (*i.e.* Katherine Brandegee) reported the specimen in "bad condition" in 1885 (*Bull. California Acad. Sci.* 1: 134). = LUPINUS CERVINUS Kell.

Hippocastanaceae

Aesculus californica Nutt. "Raised from seed sent to Exeter by William Lobb and produced fine thyrsi in July 1858" *Hortus Veitchii* p. 356, 1906. *Cf.* also *Bot. Mag.* t. 5077. 1858. = Aesculus californica (Spach) Nutt.

Rhamnaceae

Ceanothus floribundus Hook. *Bot. Mag.* t. 4806. 1854. This was based on *Lobb* 23 presumably from the vicinity of Monterey (Jepson, *A flora of California* 2: 464, 1936). According to W. J. Bean (*Trees and shrubs hardy in the British Isles* 1: 318, 1936) the only specimen at Kew is the one made in 1854 from the garden-grown plant in Veitch's nursery. *Cf.* Van Rensselaer and McMinn, *Ceanothus*, pp. 32 and 219, 1942. = Ceanothus dentatus var. floribundus (Hook.) Trel.

Ceanothus leucodermis Greene based on *Lobb* 26 from an unknown locality in California. *Cf.* Van Rensselaer and McMinn, *Ceanothus*, p. 190, 1942, for the complicated history of this species. = Ceanothus leucodermis Greene.

Ceanothus lobbianus Hook. *Bot. Mag.* t. 4811. 1854, based on Veitch garden grown plant, presumably from vicinity of Monterey where, according to Van Rensselaer (*l.c.* p. 54) the two parents of this hybrid, *C. griseus* and *C. dentatus*, grow natively. McMinn writes: "It has seldom been collected in the wild since the time of Mr. Lobb's original collection in about 1853" (*l.c.* p. 270). = Ceanothus × lobbianus (Hook.) McMinn.

Ceanothus oreganus Nutt. "Recently imported ... by Messrs. Veitch ... from the Oregon territory, through their collector, Mr. William Lobb" *Bot. Mag.* t. 5177. 1860. *Cf.* also *Hortus Veitchii* p. 394, 1906. = Ceanothus sanguineus Pursh.

Ceanothus papillosus Torr & Gray "William Lobb sent seed to Exeter ... flowered for the first time in June 1854" *Hortus Veitchii* p. 394, 1906, *Cf.* also *Bot. Mag.* t. 4815. 1854. = Ceanothus papillosus Torr. & Gray.

Ceanothus veitchianus Hook. *Bot. Mag.* t. 5127. 1859. "... collected by Mr. Lobb about 1853, apparently in the vicinity of Monterey ... Mr. Lobb undoubtedly had a keen sense of observation and a remarkable alertness for oddities, as *C. Veitchianus* has never since been found in the wild. Two other rare ceanothuses collected by him about the same time and in the same vicinity include *C. Lobbianus* and *C. floribundus* Hook." (Van Rensselaer and McMinn, *Ceanothus*, p. 97, 1942). Hybrid between *C. griseus* and probably *C. rigidus*. = Ceanothus × veitchianus (Hook.) McMinn.

Ceanothus velutinus Dougl. "Not introduced to cultivation until sent home by William Lobb" *Hortus Veitchii*, p. 394, 1906. Van Rensselaer (*Ceanothus*, p. 100, 1942) says that Lobb introduced this species "about 1853." = Ceanothus velutinus Dougl.

Sterculiaceae

Fremontia californica Torr. First grown in England "from seed sent by Mr. Robert Wrench from California ... a plant flowered for the first time in April 1854. Seed subsequently sent by William Lobb, from California." *Hortus Veitchii*, p. 367, 1906. *Cf.* also *Bot Mag.* t. 5591. 1866. = Fremontia californica Torr.

VIOLACEAE

Viola pedunculata Torr. & Gray "One of the many discoveries of Douglas ... since found by Mr. William Lobb" *Bot. Mag.* t. 5004. 1857. *Cf.* also *Hortus Veitchii* p. 438, 1906. Both Douglas and Lobb may well have collected this violet in the vicinity of Monterey where it is plentiful. = VIOLA PEDUNCULATA Torr. & Gray.

ONAGRACEAE

Epilobium pruinosum Hausskn., *Oesterr. Bot. Z.* 29: 91. 1879, based on two collections, by Thomas Bridges and William Lobb, from "California." = EPILOBIUM GLABERRIMUM Barbey.

Godetia deflexa Jepson, *Univ. California Publ. Bot.* 2: 333. 1907, based on William Lobb coll. at Kew from "sandy plains of Los Angeles in 1854(?)" = GODETIA DEFLEXA Jeps.

Oenothera bistorta var. *veitchiana* Hook., *Bot. Mag.* t. 5078. 1858 based on *William Lobb* 416, from vicinity of San Gabriel, Los Angeles Co., and introduced through Veitch in 1858. = OENOTHERA BISTORTA var VEITCHIANA Hook.

ERICACEAE

Leucothoë lobbii Hort. "Discovered by William Lobb in 1853 on the Sierra Nevada Mountains fo California, at an elevation of 5,000 ft." *Bot. Mag.* t. 6247. 1876. "Introduced and distributed under name *L. lobbii* and subsequently collected in the same locality by Miss N[ancy] J. Davis" *Hortus Veitchii*, p. 401, 1906. = LEUCOTHOE DAVISIAE Torr.

Rhododendron californicum Hook. Collected on "Mr. Lobb's first mission to [the mountains of California]" *Bot. Mag.* t. 4863. 1855. "Flowered for the first time in 1855." *Hortus Veitchii*, p. 406, 1906. = RHODODENDRON CALIFORNICUM Hook.

Rhododendron occidentale Gray "Raised from seed sent direct from California by William Lobb" *Bot. Mag.* t. 5005. 1857. *Cf.* also *Hortus Veitchii*, p. 407, 1906 = RHODODENDRON OCCIDENTALE Gray.

PRIMULACEAE

Primula suffrutescens Gray, *Proc. Amer. Acad. Arts* 7: 371. 1868, based on "California, Lobb, in herb. Hooker" and Brewer collection from Silver Mt., 10,000, Sierra Nevada. According to *Bot. Mag.* t. 8990. 1923, the Lobb specimens at Kew are dated "1857." = PRIMULA SUFFRUTESCENS Gray.

POLEMONIACEAE

Gilia californica Benth. "Mr. William Lobb sent seeds of it to Messrs. Veitch from San Bernardino, in South California" *Bot. Mag.* t. 4872. 1855. According to *Hortus Veitchii*, p. 421, 1906, this introduction flowered in June 1855. *Cf.* also *Gard. Chron.* 16: 423. 1855. = LEPTODACTYLON CALIFORNICUM Hook. & Arn.

Gilia dianthoides Endl. "Introduced by William Lobb. Flowered in summer of 1855." *Hortus Veitchii*, p. 421. 1906. *Cf.* also *Bot. Mag.* t. 4876. 1855. = LINANTHUS DIANTHIFLORUS (Benth.) Greene.

Gilia lutea Steud. "Messrs. Veitch received the seed in 1852 from Mr. William Lobb" *Bot. Mag.* t. 4735. 1853. *Cf.* also *Hortus Veitchii*, p. 421, 1906. = LINANTHUS ANDROSACEUS ssp. LUTEUS (Benth.) Mason.

HYDROPHYLLACEAE

Nama systylum Gray, *Proc. Amer. Acad. Arts* 6: 37. 1866, based on *Lobb* 164. Eastwood (*Leafl. W. Bot.* 5: 155. 1949) suggests Donner or Carson passes as the type locality... DRAPERIA SYSTLA (Gray) Torr.

Nama lobbii Gray, *Proc. Amer. Acad. Arts* 6: 37. 1866, based on *Lobb* 108 taken "undoubtedly in the Sierra Nevada, most probably in Placer Co. or Nevada Co." (Jepson, *A flora of California* 3: 288. 1943). Eastwood (*Leafl. W. Bot.* 5: 155. 1949) suggests Donner or Carson passes as well travelled roads to the Nevada mines. = NAMA LOBBII Gray.

Phacelia campanularia Gray "A native of San Bernardino and San Diego counties" *Bot. Mag.* t. 6735. 1884. It is likely in view of the restricted distribution of this species in the western arm of Colorado Desert, of present Riverside Co., that Lobb made a trip through San Gorgonio Pass from San Bernardino and into Marongo Valley. *Chilopsis linearis* almost certainly was taken on the same trip on which he collected this striking *Phacelia*. = PHACELIA CAMPANULARIA Gray.

Phacelia whitlavia Gray "Introduced by Lobb and flowered at the Chiswick Exhibition in July 1854" *Hortus Veitchii*, p. 428, 1906. *Cf.* also *Bot. Mag.* t. 4813. 1854. = PHACELIA MINOR (Harv.) Thell.

LABIATAE

Acanthomintha lanceolata Curran. Evidently the first collection of this species was made by Lobb at San Antonio, that is, vicinity of San Antonio de Padua Mission, Monterey Co., where David Douglas took lodging April 27, 1831. It may prove that Lobb, like Douglas and Coulter before him, took advantage of the missions as hostelries during their overland jornadas through California. Mission records have yet to be searched for more information on this head. A full documented history of the missions with special reference to Soledad may be found in Elliot A. P. Evans, "Soledad" (*J. Soc. Architect. Historians* 15(1): 20–26. 1956). This Lobb collection reported by T. S. Brandegee (*Zoe* 4: 157. 1893) was probably made on the same trip to the Santa Lucia Mts. in the spring of 1850 that Lobb collected *Abies bracteata*. = ACANTHOMINTHA LANCEOLATA Curran.

SCROPHULARIACEAE

Collinsia bartsiaefolia Benth. "Distributed in 1852 from seed sent by Lobb" *Hortus Veitchii*, p. 417. 1906 = COLLINSIA BARTSIAEFOLIA Benth.

Collinsia multicolor Lindl. "Distributed in 1852, sent by Lobb." *Hortus Veitchii*, p. 417. *Cf. Gard. Chron.* 13: 690. 1852 = COLLINSIA HETEROPHYLLA Buist ex Grah.

Mimulus brevipes Benth. "W. R. Maxon and I [W. L. Jepson] went one day to the Lindley Herbarium in the Botany School [at Cambridge University]. *Mimulus brevipes*, a sp[eci]m[en] by W. Lobb 400, is marked "mts. of San Gabriel" "annual 2-4 ft. high." so Lobb did occasionally give a station for a collection —" Aug. 21, 1930, Jepson, Field Book, ms. 51:

p. 94. Univ. Calif., Berkeley. For a mention of the Lobb collections at Cambridge see Gilmour and Tutin, 1933, p. 22. = MIMULUS BREVIPES Benth.

Penstemon lobbii Hort. ex Lemaire, *Ill. Hort.* t. 315. 1862. It is noteworthy that Hooker (*Bot. Mag.* t. 6157. 1875) makes no reference to Lobb's introduction. = PENSTEMON ANTIRRHINOIDES Benth.

Penstemon jaffrayanus Hook. *Bot. Mag.* t. 5045. 1858, based on plant grown from seed "received from Mr. William Lobb" which evidently flowered in 1857 according to *Hortus Veitchii*, p. 428. 1906. = PENSTEMON PARVULUS (Gray) Krautter.

BIGNONIACEAE

Chilopsis linearis DC. "California (W. Lobb! Coulter, no. 634!)" B. Seemann, *The botany of the voyage of H.M.S. Herald*, p. 326. 1857. Lobb must have taken *Chilopsis* on this trip to Morongo Valley, Riverside Co., in 1854. = CHILOPSIS LINEARIS (Cav.) Sweet.

Hemizonia lobbii Greene, *Bull. Torrey Bot. Club* 9: 109. 1882, based "on a single specimen, coll. by Lobb, thought to come from near Monterey" (A. Gray, *Synoptical flora of North America* 1 (pt. 2): 310. 1884). For a basis of separating *H. lobbii* from *H. fasciculata* see D. D. Keck, *Madroño* 3: 8. 1935. = HEMIZONIA LOBBII Greene.

Finding List of Numbered California Collections

It was hoped that listing Lobb's numbered specimens would elucidate his California itineraries. His numbers, however, were evidently not given in the field, but at the time of their shipment to Veitch. For example, his numbers 95 and 119 are certainly from the Santa Lucia Mountains, but within that sequence falls no. 108, which could not have been collected in that range.

This list is based primarily on a set of specimen at the British Museum (Natural History), prepared with the assistance of Eric Groves, with the identifications of Peter Raven which have been verified or, in a few instances, changed, by the author. Several specimens are numbered identically, further evidence that they were not numbered in the field. In those rare instances where localities appear on the labels, they are quoted here. A few specimens, separately designated, are not represented in the British Museum (Natural History) set: Cambridge University (CGE), Gray Herbarium (GH), and Kew (K).

1. *Rhododendron macrophyllum*
2. *Rosa californica*
4. *Eriogonum parvifolium*
5. *Adenostoma fasciculatum* "Monterey"
6. *Ceanothus papillosus*
8. *Photinia arbutifolia*
13. *Rhamnus californica*
21. *Quercus agrifolia*
23. *Ceanothus floribundus* Type coll.
24. *Ceanothus integerrimus*
25. *Ceanothus cuneatus*
26. *Quercus chrysolepis*
26. *Ceanothus leucodermis* Type coll.
27. *Ceanothus papillosus*
28. *Quercus wislizeni*
30. *Viola douglasii*
32. *Oenothera deltoides*
33. *Oenothera dentata*
39. *Silene laciniata*
46. *Cynoglossum grande*
48. *Erysimum capitatum*
58. *Trientalis latifolia*
60. *Thysanocarpus radians*
64. *Chilopsis linearis*
66. *Limnanthes douglasii*
67. *Limnanthes alba*
74. *Orthocarpus erianthus*
74. *Penstemon breviflorus*
79. *Chilopsis linearis*
82. *Calycadenia multiglandulosa*
83. *Navarettia viscidula*
84. *Navarettia leucocephala*
85. *Navarretia ? pubescens*
90. *Castelleja foliolosa*
95. *Dendromecon rigidum* "summit of Santa Lucia range"
96. *Salvia spathacea*
99. *Centaurium* sp.
103. *Quercus wislizenii*
104. *Quercus lobata*
108. *Nama lobbii* Type coll.
109. *Quercus chrysolepis*
110. *Lithocarpus densiflora*
112. *Lotus scoparius*
115. *Aconitum columbianum*
116. *Umbellularia californica*
118. *Kelloggia galioides*
119. *Lupinus cervinus* Type coll. "Santa Lucia Mts."
120. *Campanula prenanthoides*
134. *Diplacus aurantiacus* "Santa Lucia Mts."
139. *Gilia aggregata*
142. *Penstemon* sp.
142. *Collomia heterophylla*
142. *Navarretia pubescens*
147. *Arctostaphylos hookeri*
149. *Cornus nuttallii*
151. *Linanthus androsaceus*
152. *Ceanothus cuneatus*
153. *Linanthus androsaceus*
153. *Viola sempervirens*
156. *Viola beckwithii*
157. *Rhododendron occidentale*
158. *Geum macrophyllum*
158. *Oxalis oregana*
159. *Geranium richardsonii*
159. *Abronia latifolia*

160. *Delphinium cardinale*
160. *Cycladenia humilis*
163. *Oenothera ovata*
164. *Nama systylum* Type coll.
164. *Scutellaria antirrhinoides*
166. *Frasera albicaulis*
168. *Gentiana oregana*
170. *Trichostema oblongum*
171. *Spiraea lucida*
172. *Physocarpus capitatus*
174. *Lonicera involucrata*
175. *Verbena scabra*
176. *Berberis aquifolium*
178. *Rhamnus californica*
182. *Prunus emarginata*
183. *Prunus demissa*
184. *Crataegus douglasii*
185. *Pycnanthemum californicum*
188. *Lonicera hispidula* var. *vacillans*
190. *Eriogonum lobbii* Type coll.
192. *Anemone quinquefolia* var. *grayi*
193. *Eriogonum umbellatum*
195. *Ceanothus integerrimus*
199. *Silene californica*
200. *Asarum caudatum*
201. *Dentaria californica*
202. *Dicentra formosa*
212. *Viburnum* sp.
214. *Ribes californicum*
217. *Rosa gymnocarpa*
217. *Platanthera leucostachya*
231. *Micromeria douglasii*
242. *Calochortus maweanus*
243. *Calochortus minimus*
249. *Ranunculus hydrocharis* f. *lobbii* Type coll.
257. *Calochortus elegans* var. *lobbii*
264. *Lupinus aridus* var. *lobbii* Type coll.
277. *Polygala cornuta*
301. *Sarcodes sanguinea*
311. *Cercis occidentalis*
312. *Castanopsis chrysophylla*
313. *Aesculus californica*
315. *Fremontia californica*
318. *Rhus diversiloba*
323. *Vitis californica*
325. *Cornus nuttallii*
326. *Cornus glabrata*
328. *Acer negundo*
329. *Acer glabrum*
330. *Rubus parviflorus*
332. *Photinia arbutifolia*
339. *Rhododendron macryphyllum*
340. *Holodiscus discolor*
341. *Amelanchier pallida*
342. *Physocarpus capitatus*
344. *Holodiscus dumosus* ssp. *saxicola*
345. *Styrax californica*
346. *Umbellularia californica*
349. *Populus trichocarpa*
352. *Philadelphus californicus*
359. *Berberis aquifolium*
360. *Photinia arbutifolia*
361. *Prunus ilicifolia*
370. *Ceanothus velutinus*
373. *Ceanothus oliganthus*
375. *Ceanothus sorediatus*
376. *Arctostaphylos tomentosa*
385. *Encelia californica*
386. *Helianthus californicus*
387. *Helianthus gracilentus*
388. *Mirabilis laevis*
389. *Phacelia grandiflora*
390. *Phacelia campanularia*
392. *Phacelia viscida*
395. *Leptodactylon californicum*
398. *Penstemon centranthifolius*
400. *Mimulus brevipes* "Mts. of San Gabriel"
410. *Fremontia californica*
411. *Oenothera* sp. "plains of San Gabriel"
416. *Oenothera bistorta* var. *veitchiana* "San Gabriel" Type coll.
418. *Delphinium cardinale*
426. *Stylomecon heterophylla*
434. *Penstemon cordifolius*
457. *Ceanothus oliganthus*
462. *Phacelia cicutaria* var. *hispida*
468. *Centaurium venustum*
471. *Gutierrezia bracteata*
472. *Lepidospartum squamatum*
475. *Eriophyllum* sp.
476. *Eriophyllum confertiflorum*
507. *Lupinus nanus*
513. *Galium angustifolium*
513. *Vitis californica*
533. *Chaetopappa aurea*
535. *Malacothrix californica*
541. *Delphinium nudicaule*
542. *Penstemon* sp.
545. *Styrax californica*
546. *Ceanothus oliganthus*
548. *Phlox speciosa* ssp. *occidentalis*
551. *Lotus grandiflorus*
552. *Brodiaea ida-maia*

The following species have unnumbered labels:

Cupressus macrocarpa
Frasera parryi
Chamaebatia foliolosa
Gilia capitata
Isopyrum occidentale "Sierra Nevada. 6000 ft."
Paeonia californica "Santa Lucia Mts. 3000 ft."

Acknowledgments

In the preparation of this account I am especially grateful to Sir George Taylor, formerly Director, Royal Botanic Gardens, Kew, for permission to quote from the Hooker letters; and to Raymond Desmond, formerly Librarian, Kew, for his several acts of assistance in their study. To Eric Groves, British Museum (Natural History); Dr. Rimo Bacigalupi, University of California, Berkeley; Dr. Elliot A. P. Evans, Society of California Pioneers, San Francisco; and Mrs. Lazella Schwarten, formerly Librarian, Arnold Arboretum, for their various favors. Mrs. Ruth Hunkin, Oxford, England, has taken a keen interest in the search for Lobb materials. Dr. George Lindsay, Director, California Academy of Sciences, and Ray Brian, Librarian, were essential in the search of the Academy's Minutes. I am grateful to Dr. Lindsay for permission to quote from them.

Again, and on two continents, I have had the close and esoteric assistance of my wife, Nesta Dunn Ewan, in the progress of this sketch.

Dr. T. S. Palmer, who answered my letters of inquiry regarding Lobb in 1935, and many others, cannot accept my thanks today but my acknowledgment is none-the-less sincere.

References

ANON. — "The big trees of California." *Harper's weekly* 2: 357–358. 1858.
Lobb not mentioned. Impressive illustration.

ANON. — Short account of the big trees of California. *U. S. Dept. Agric. Div. of Forestry Bull.* 28: 1–30. 1900.
Lobb not mentioned. A compilation made with assistance of William Russell Dudley.

BAY, JENS CHRISTIAN BALLIEN—Sequoia and Washingtonia In History and Bibliography. *Northwestern Univ. Med. School Quart. Bull.* 20(1): 13–17. 1946. Also issued as a repaged reprint.
Washingtonia gigantea herein proposed as a new combination.

BIGELOW, JOHN MILTON — Description of Forest Trees. In *Pacific railroad reports of explorations and surveys* ... Washington, D. C., Senate Document, 1856, Vol. IV, Part V, No. 2: "Wellingtonia gigantea," pp. 22–23.
Issued Sept., 1857, *teste* I. M. Johnston, *J. Arnold Arbor.* 24: 241. 1943. Bigelow confused Lobb's sponsor with Jeffrey's and so gave him a Scottish connection. Jepson's date for the communication of Big Tree specimens to Torrey and Gray in 1852 is here.

BRITTEN, JAMES & BOULGER, GEORGE SIMONDS — *Biographical index of deceased British and Irish botanists.* Ed. 2, revised and completed by Alfred Barton Rendle. London, Taylor and Francis, 1931.

CANDOLLE, ALPHONSE DE — *La phytographie* ... Paris, G. Masson, 1880.
de Candolle states, p. 429, that Lobb duplicates are at the University of Dublin, and in the Delessert collection, now at Conservatorei Botanique, Geneva.

CLOKIE, HERMIA NEWMAN — *An account of the herbaria of the department of botany in the University of Oxford.* Oxford, Oxford Univ. Press, 1964.
Three Lobb "California" specimens, perhaps acquired by Fielding, mentioned p. 202.

COVILLE, FREDERICK VERNON — Itinerary of John Jeffrey, an early botanical explorer of western North America. *Proc. Biol. Soc. Wash.* 11: 57-60. 1897.

DALLIMORE, WILLIAM — William Lobb & John Jeffrey. *Forestry (Oxford)* 6: 5-8. 1932.

DAYTON, WILLIAM ADAMS — Names of the Giant Sequoia. *Leafl. W. Bot.* 3: 209-219. 1943.

EASTWOOD, ALICE — Explorations of William Lobb. *Muhlenbergia* 7: 100-103. 1911.
Principally a republication of Veitch's account of 1906.

——— — Early botanical explorers on the Pacific Coast and the trees they found there. *Quart. California Hist. Soc.* 18(4): 335-346. 1939.
Lobb account on p. 343. Death year persistently misstated as 1863 instead of 1864. Reprinted in Susanna Bryant Dakin, *Perennial adventure. A tribute to Alice Eastwood, 1859-1953.* San Francisco, Calif. Acad. Sci., 1954, 23-40, 43-48; Lobb on pp. 37-38.

ELLSWORTH, RODNEY Sydes — *The giant sequoia. An Account of the history and characteristics of the Big Trees of California.* Oakland, J. D. Berger, 1924.
Lobb not mentioned. Useful bibliography, pp. 159-167.

——— — Discovery of the Big Trees of California, 1833-1852, Being a study of the various claims of discovery of the Giant Sequoia of the Sierra Nevada, with new light on the discovery of the Calaveras Grove in 1850 by John Marshall Wooster based on original material. Univ. California, M.A. Thesis, Dec. 1933, *ined.*
Extension of earlier published account. Carbon copy at Bancroft Library, University of California.

FARQUHAR, FRANCIS PELOUBET — *Yosemite, the Big Trees, and the High Sierra.* Berkeley & Los Angeles, Univ. of California Press, 1948.

FRITZ, EMANUEL — *California coast redwood, an annotated bibliography to and including 1955.* San Francisco, Foundation for American Resource Development, 1957.

FRY, WALTER & WHITE, JOHN ROBERTS — *Big Trees.* Palo Alto, Stanford Univ. Press, 1930, and various printings; [ed. 2] "8th" printing, 1946.
"Standard" account, but does not mention Lobb.

GILMOUR, JOHN SCOTT LENNOX & TUTIN, THOMAS GASKELL — *A list of the more important collections in the university herbarium Cambridge.* Cambridge, Univ. Press, 1933.
Lobb collections acquired by Lindley, p. 22.

GORDON, GEORGE & GLENDINNING, ROBERT — *The pinetum: being a synopsis of all the coniferous plants at present known* ... London, Henry G. Bohn, 1858.
Interesting that Lobb's name is not mentioned under *Wellingtonia; Thuja lobbii,* etc., although from the preface we learn that Gordon was then communicating with Mr. Low of Clapton Nursery.

GORDON, GEORGE — *A supplement to Gordon's Pinetum* ... London, Henry G. Bohn, 1862.
Several references to Lobb as collector, not for Veitch, but for Clapton Nursery.

——— — *The pinetum* ... Ed. 2. London, Henry G. Bohn, 1875.
Essentially same text as *Supplement,* but with some persistent errors such as crediting Douglas with the discovery of the Big Tree in 1831.

GORDON-CUMMING, CONSTANCE FREDERICA — *Granite crags.* Edinburgh & London, William Blackwood & Sons, 1884.

Noteworthy among an extensive travelers' literature on Big Trees. Her account, originally written as letters home, includes many natural history notes, often with scientific names, as well as social impressions. Precise figures on the comparative height of Big Tree and Australian *Eucalyptus* are given on p. 310. Writings of Muir, Whitney, etc., are cited, but Lobb is not mentioned. Farquhar, p. 69-71, provides bibliographic notes on this book.

HADFIELD, MILES — *Gardening in Britain.* London, Hutchinson & Co., 1960.
The standard history of British horticulture. Short account of Lobb on p. 323.

HART, ANN CLARK — *Lone mountain. The most revered of San Francisco's hills. Annals of the pioneers copied from headstones and other old records.* San Francisco, Pioneer Press(?), 1937.
No mention of Lobb among the selection of 47,000 names on the grave books of Laurel Hill, first dedicated in 1854 as Lone Mountain Cemetery.

[HEMSLEY, WILLIAM BOTTING] — *Official guide to the North Gallery, Royal Gardens, Kew.* Ed. 6. London, H.M. Stationery Office, 1914.
In this catalogue of the paintings of Marianne North (1830-90), her "Calaveras Grove of the Big Tree, or Wellingtonia, in the evening," is described on p. 24. Lobb did not "rediscover" the grove in 1853, as there stated. Miss North does not elaborate on her visit in her *Recollections of a happy life* (2 vols. London & New York, Macmillan & Co.) 1: 208-209. 1892.

HOOPES, JOSIAH — *Book of evergreens.* New York, Orange Judd, 1868.
"The English naturalist Lobb is supposed by many to have first met with [the Big Tree] near the source of the river Stanislaus" (p. 241), but Lobb actually collected his cones, etc., in Calaveras Grove. This often overlooked handbook, with excellent woodcuts, has many historical notes on California conifers supplied especially by Parry, Brewer, and Meehan.

HUNKIN, JOSEPH WELLINGTON — William and Thomas Lobb: two Cornish Plant Collectors. *J. Roy. Hort. Soc.* 67: 48-51. (Feb.) 1942.

——— — A further note on the brothers William and Thomas Lobb "by the Rt. Rev. the Lord Bishop of Truro [J. W. Hunkin, D.D.]." *J. Roy. Hort. Soc.* 72: 33-35. 1947.

JEPSON, WILLIS LINN — Thomas Bridges. *Madroño* 2: 84-88. 1933.
Bridges merely mentions Lobb in a letter, San Francisco, May 5, 1858, to W. J. Hooker, telling of his own plan to explore the Sierra Nevada summits for plants.

——— — Albert Kellogg. In *Dictionary of American biography* 10: 300. 1946.

JOHNSTON, IVAN MURRAY — Botanical activities of Thomas Bridges. *Contr. Gray Herb.* no. 81: 98-106. 1928.

JOLY, CHARLES — Note sur les arbres géants de la Californie. *J. Soc. Natl. Hort. France,* Sér. 3. 5: 185-193. 1883.
London Athenaeum and *Gardener's chronicle* articles mentioned, but no mention of Lobb. Two specimens of Big Tree presented by M. Decaisne.

KELLOGG, ALBERT — *Forest trees of California.* [Sacramento, 1882]. Appendix to State Mineralogist's Reports, State Mining Bureau, Henry G. Hanks, State Mineralogist.
Lobb mentioned in footnote on p. 19.

KENT, ADOLPHUS HENRY — *Veitch's manual of the Coniferae.* New ed. London, James Veitch & Sons, 1900.

KILLIP, ELLSWORTH PAINE — Botanical collections of William Lobb in Colombia. *Smithsonian Misc. Collect.* 87: 1-13. 1932.

LEMMON, JOHN GILL — Cone-bearers of California. In *Third Biennial Report of the California State Board of Forestry for the Years 1889-90.* Sacramento, State Printing Office, 1890, p. 75-212.
Lobb not mentioned.

——— — *Handbook of West-American cone-bearers.* Ed. 4. Oakland, The Author, 1900.
Lobb referred to only indirectly: "Discovered in 1852, specimens [of the Big Tree] were sent in 1853 to Dr. Lindley ...," p. 109.

LEONARD, ZENAS — *Adventures of Zenas Leonard, fur trader.* Ed. by John Canfield Ewers Norman, Univ. Oklahoma Press, 1959.
Notice of the Big Tree on p. 83. Another ed. edited by William Fremont Wagner, Cleveland, Burrows Bros., 1904, notices Big Trees on p. 179-180.

LINDLEY, JOHN — Wellingtonia gigantea. *Gard. Chron.* 13: 819-820, 823. (Dec. 24) 1853.

LITTLE, ELBERT LUTHER, JR. — *Check list of native and naturalized trees of the United States (including Alaska).* U. S. Dept. Agric. Handbook no 41. Washington, Forest Service, 1953.

[LOUDON, JOHN CLAUDIUS, (ed.)]—[Report on] Provincial horticultural societies, Cornwall. *Gard. Mag.* (Loudon) 13: 552. 1837.

McKELVEY, SUSAN DELANO — *Botanical exploration of the Trans-Mississippi West, 1790-1850* Jamaica Plain, Mass., Arnold Arboretum of Harvard University, 1955.
Lobb account on p. 1092-1094.

MARSHALL, WALTER GORE — *Through America: or, nine months in the United States.* London, S. Low, Marston, Searle & Rivington, 1881.
Lobb mentioned on p. 332. Relates Wooster's sighting of the Big Tree in 1850.

[NELSON, JOHN] — *Pinaceae: being a handbook of the firs and pines* "by Senilis *[pseud.]*" London, Hatchard & Co., 1866.
Exhibition of the Big Tree in the Crystal Palace, London, is told on p. 80-81. *Gigantabies* proposed on p. 77.

PARRY, CHARLES CHRISTOPHER — Early botanical explorers of the Pacific coast. *Overland Monthly* ser. 22. 2: 415-416. 1883.
Separately paged reprint in Brandegee Collection, University of California, Berkeley.

REHDER, ALFRED — *Bibliography of cultivated trees and shrubs hardy in the cooler temperate regions of the Northern Hemisphere.* Jamaica Plain, Mass., Arnold Arboretum of Harvard University, 1949.

RODGERS, ANDREW DENNY, III– *John Torrey, a story of North American botany.* Princeton, University Press, 1942.
Lobb's excursion with Parry to "Mission of San Juan" mentioned on p. 222. [— Mission of San Juan Bautista, Monterey Co.]

SARGENT, CHARLES SPRAGUE — *Silva of North America* ... 14 vols. Boston & New York, Houghton, Mifflin & Co.
Lobb biographical note in vol. 10, p. 80; Big Tree note on p. 147.

SEEMANN, BERTHOLD CARL — On the Mammoth Tree of upper California. *Ann. Mag. Nat. Hist.* ser. 3. 3: 161. 1859.
Lobb not mentioned.

SENILIS — see [Nelson, John].

SHINN, CHARLES HOWARD — The great sequoia. *Gard. & Forest* 2: 614-615. 1889.
General Bidwell is credited with its discovery in 1841. Shinn follows Bigelow's account of Lobb.

T[___], G[___] W[___] — William Lobb. *Gard. Chron.* ser. 3. 104: 388-389. (Nov. 26) 1938.
No information given here that is not more fully available elsewhere. Fig. 167, p. 389, photograph of Carberry Tower Garden shows a fine Big Tree "which must be without doubt [one of] the finest specimens in the country."

URBAN, IGNATZ—William Lobb. *In* Karl Friederich Philipp von Martius, *Flora Brasiliensis.* Stuttgart & Tubigen. 1(1): 46. 1906.
Summarizes Lobb's Brazilian travels, with biographical references.

VAN RENSSELAER, MAUNSELL & MCMINN, HOWARD EARNEST—*Ceanothus.* Part I (Van Rensselaer), *Ceanothus for gardens, parks, and roadsides.* Part II (McMinn), *A systematic study of the genus Ceanothus.* Santa Barbara, Santa Barbara Botanic Garden, 1942.

VEITCH, JAMES HERBERT—*Hortus Veitchii, A history of the rise and progress of the nurseries of Messrs. James Veitch and Sons, together with an account of the botanical collectors ... and a list of the most remarkable of their introductions.* London, James Veitch & Sons, Ltd., 1906.
Biographical account of Lobb on pp. 37-40.

WHITNEY, JOSIAH DWIGHT—*Yosemite Guide-book* ... [Sacramento,] Published by the authority of the Legislature, 1869.
Account of Big Tree is based on Brewer and others, pp. 139-155.

Addendum

Two titles have come to my notice since this paper was submitted for publication:

ELAND, G.—*The Lobb Family from the Sixteenth Century.* Privately printed for Yda Cory-Wright, University Press, Oxford, 1955.
Early history of Cornwall Lobbs told but William Lobb (1809–1864) is not mentioned.

TAYLOR, NORMAN—*The Ageless Relicts. The Story of Sequoia.*
St. Martin's Press, New York, 1962.
Lobb is noticed in this undocumented but useful, rather too evangelical, story.

Index of Plant Names

Collection numbers in parentheses ( ).

Abies amabilis 20
Abies bracteata 5, 20
Abies concolor 20
Abies concolor var. *lowiana* 20, 21
Abies grandis 20
Abies magnifica 20
Abies nobilis 20
Abies pattoni 20
Abies procera 20
Abronia latifolia (159) 25
Acanthomintha lanceolata 5, 26
Acer glabrum (329) 29
Acer negundo (328) 29
Aconitum columbianum (115) 28
Adenostoma fasciculatum (5) 28
Aesculus californica (313) 24, 29
Amelanchier pallida (341) 29
Anemone quinquefolia var. *grayi* (192) 29
Arctostaphylos hookeri (147) 28
Arctostaphylos tomentosa (376) 29
Asarum caudatum (200) 29

Berberis aquifolium (176, 359) 29
Brodiaea ida-maia (552) 29

Calochortus caeruleus 21
Calochortus lobbii 21
Calochortus elegans 21, 29
Calochortus maweanus 21, 29
Calochortus minimus 22, 29
Calochortus tolmei 21
Calycadenia multiglandulosa (82) 28
Campanula prenanthoides (120) 28
Castanopsis chrysophylla (312) 22, 29
Castilleia foliolosa (90) 28
Ceanothus cuneatus (25, 152) 28
Ceanothus dentatus var. *floribundus* 24
Ceanothus floribundus (23) 24, 28
Ceanothus griseus × *rigidus* 24
Ceanothus integerrimus (24, 195) 28, 29
Ceanothus leucodermis (26) 24, 28
Ceanothus lobbianus 24
Ceanothus oliganthus (373, 457, 546) 29
Ceanothus oreganus 24
Ceanothus papillosus (6, 27) 24, 28
Ceanothus rigidus 24
Ceanothus sanguineus 24
Ceanothus serediatus (375) 29
Ceanothus veitchianus 24
Ceanothus velutinus (370) 24, 29
Centaurium sp. (99) 28
Centaurium venustum (468) 29
Centrostegia leptoceros 22
Cercis occidentalis (311) 29
Chaetopappa aurea (533) 29
Chamaebatia foliolosa 23, 29
Chilopsis lineuris (64, 79) 3, 10, 26, 27, 28

Chorizanthe leptoceras 22
Collinsia bartsiaefolia 26
Collinsia heterophylla 26
Collinsia multicolor 26
Collomia heterophylla (142) 28
Cornus glabrata (326) 29
Cornus nuttallii (149, 325) 28, 29
Crataegus douglasii (184) 29
Cupressus macrocarpa 29
Cycladenia humilis (160) 29
Cynoglossum grande (46) 28

Delphinium cardinale (160, 418) 3, 22, 29
Delphinium nudicaule (541) 29
Dendromecon rigidum (95) 23, 28
Dentaria californica (201) 29
Dicentra chrysantha 23
Dicentra formosa (202) 29
Diplacus aurantiacus (134) 28
Draperia systyla 26

Encelia californica (385) 29
Epilobium glaberrimum 25
Epilobium pruinosum 25
Eriogonum lobbii (190) 14, 22, 29
Eriogonum parvifolium (4) 28
Eriogonum umbellatum (193) 29
Eriophyllum sp. (475) 29
Eriophyllum confertiflorum (476) 29
Erysimum capitatum (48) 28
Eschscholtzia lobbii 23
Eschscholtzia tenuifolia 23

Frasera albicaulis (166) 29
Frasera parryi 29
Fremontia californica (315, 410) 24, 29

Galium angustifolium (513) 29
Gentiana oregana (168) 29
Geranium richardsonii (159) 28
Geum macrophyllum (158) 28
Gilia aggregata (139) 28
Gilia californica 25
Gilia dianthoides 25
Gilia lutea 26
Godetia deflexa 25
Gutierrezia bracteata (471) 29

Helianthus californicus (386) 29
Helianthus gracilentus (387) 29
Hemizonia lobbii 27
Holodiscus discolor (340) 29
Holodiscus dumosus ssp. *saxicola* (344) 29

Iris douglasiana 22
Isopyrum occidentale 29

Juniperus californica (125) 21
Juniperus pyriformis 21

Kelloggia galioides (118) 28

Lepidospartum squamatum (472) 29
Leptodactylon californicum (395) 25, 29
Leucothoe davisiae 25
Leucothoe lobbii 25
Limnanthes alba (67) 28
Limnanthes douglasii (66) 28
Linanthus androsaceus (151, 153) 26, 28
Linanthus dianthiflorus 25
Lithocarpus densiflora (110) 28
Lithospermum ruderale (540) 29
Lonicera hispidula var. *vacillans* (188) 29
Lonicera involucrata (174) 29
Lotus grandiflorus (551) 29
Lotus scoparius (112) 28
Lupinus aridus var. *lobbii* 23, 29
Lupinus cervinus (119) 23, 28
Lupinus lobbii 23
Lupinus nanus (507) 29

Malacothrix californica (535) 29
Mimulus brevipes (400) 26, 29
Mirabilis laevis (338) 29
Micromeria douglasii (231) 29

Nama lobbii 14, 26, 28
Nama systyla 26, 29
Navarretia leucocephala (84) 28
Navarretia pubescens (85, 144) 28
Navarretia viscidula (83) 28

Oenothera sp. (411) 29
Oenothera bistorta var. *veitchiana* (416) 25, 29
Oenothera deltoides (32) 28
Oenothera dentata (33) 28
Oenothera ovata (163) 29
Orthocarpus erianthus (74) 28
Oxalis oregana (158) 28

Paeonia californica 29
Penstemon sp. (542) 28, 29
Penstemon antirrhinoides 12, 27
Penstemon brevifolius (74) 28
Penstemon centranthifolius (398) 29
Penstemon cordifolius (434) 29
Penstemon jeffreyanus 27
Penstemon lobbii 27
Penstemon parvulus 27
Phacelia campanularia (390) 3, 26, 29
Phacelia cicutaria var. *hispida* (462) 29
Phacelia grandiflora (389) 29
Phacelia minor 26
Phacelia viscida (392) 29
Phacelia whitlavia 26
Philadelphus californica (352) 29
Phlox speciosa ssp. *occidentalis* (548) 29
Phlox speciosa ssp. *occidentalis* (548) 29
Photinia arbutifolia (8, 332, 360) 28, 29
Physocarpus capitatus (172, 342) 29
Picea grandis 20
Picea lowiana 20
Pinus lophosperma 13, 21
Pinus torreyana 21
Platanthera leucostachys (217) 29
Polygala cornuta (277) 29
Populus trichocarpa (349) 29
Primula suffrutescens 25
Prunus demissa (183) 29
Prunus emarginata (182) 29
Prunus ilicifolia (361) 29
Pycnanthemum californicum (185) 29

Quercus agrifolia (21) 28
Quercus chrysolepis (26, 109) 28
Quercus lobata (104) 28
Quercus wislizenii (28, 103) 28

Ranunculus hydrocharis f. *lobbii* (249) 22, 29
Ranunculus lobbii (249) 22
Rhamnus californica (13, 178) 28, 29
Rhododendron californicum 25
Rhododendron macrophyllum (1, 339) 28, 29
Rhododendron occidentale (157) 25, 28
Rhus diversiloba (318) 29
Ribes californicum (214) 29
Ribes lobbii 23
Rosa californica (2) 28
Rosa gymnocarpa (217) 29
Rubus parviflorus (330) 29

Salvia spathacea (96) 28
Sarcodes sanguinea (301) 29
Scutellaria antirrhinoides (164) 29
Sequoia gigantea 17, 21
Sequoiadendron giganteum 3, 17, 21
Silene californica (199) 29
Silene laciniata (39) 28
Spiraea lucida (171) 29
Spraguea umbellata 22
Stylomecon heterophylla (426) 29
Styrax californica (345, 545) 29

Thuja lobbii 21
Thuja plicata 21
Thysanocarpus radians (60) 28
Torreya californica 21
Torreya myristica 21
Trichostema oblongum (170) 29
Trientalis latifolia (58) 28
Tsuga mertensiana 20

Umbellularia californica (116, 346) 28, 29

Verbena scabra (175) 29
Viburnum sp. (212) 29
Viola beckwithii (156) 28
Viola douglasii (30) 28
Viola pedunculata 25
Viola sempervirens (153) 28
Vitis californica (323, 513) 29

Wellingtonia gigantea 3, 7, 8, 17, 21

PLATES

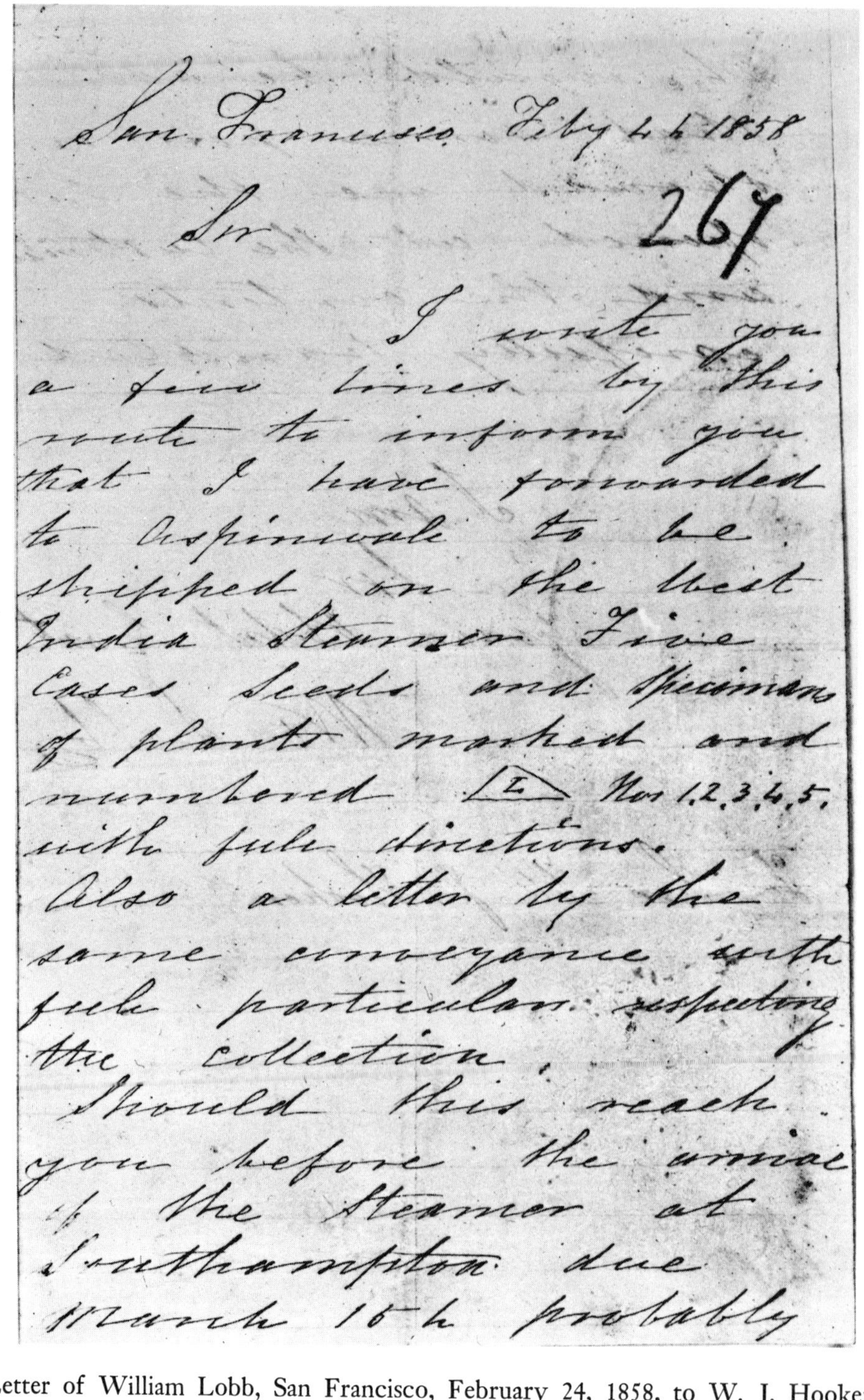

San Francisco Feby 24 1858

267

Sir

I write you a few lines by this route to inform you that I have forwarded to Aspinwale to be shipped on the West India Steamer Five Cases Seeds and Specimens of plants marked and numbered L Nos 1.2.3.4.5. with full directions. Also a letter by the same conveyance with full particulars respecting the collection. Should this reach you before the arrival of the Steamer at Southampton due March 16th probably

Letter of William Lobb, San Francisco, February 24, 1858, to W. J. Hooker (Hooker Correspondence, Royal Botanic Garden, Kew).

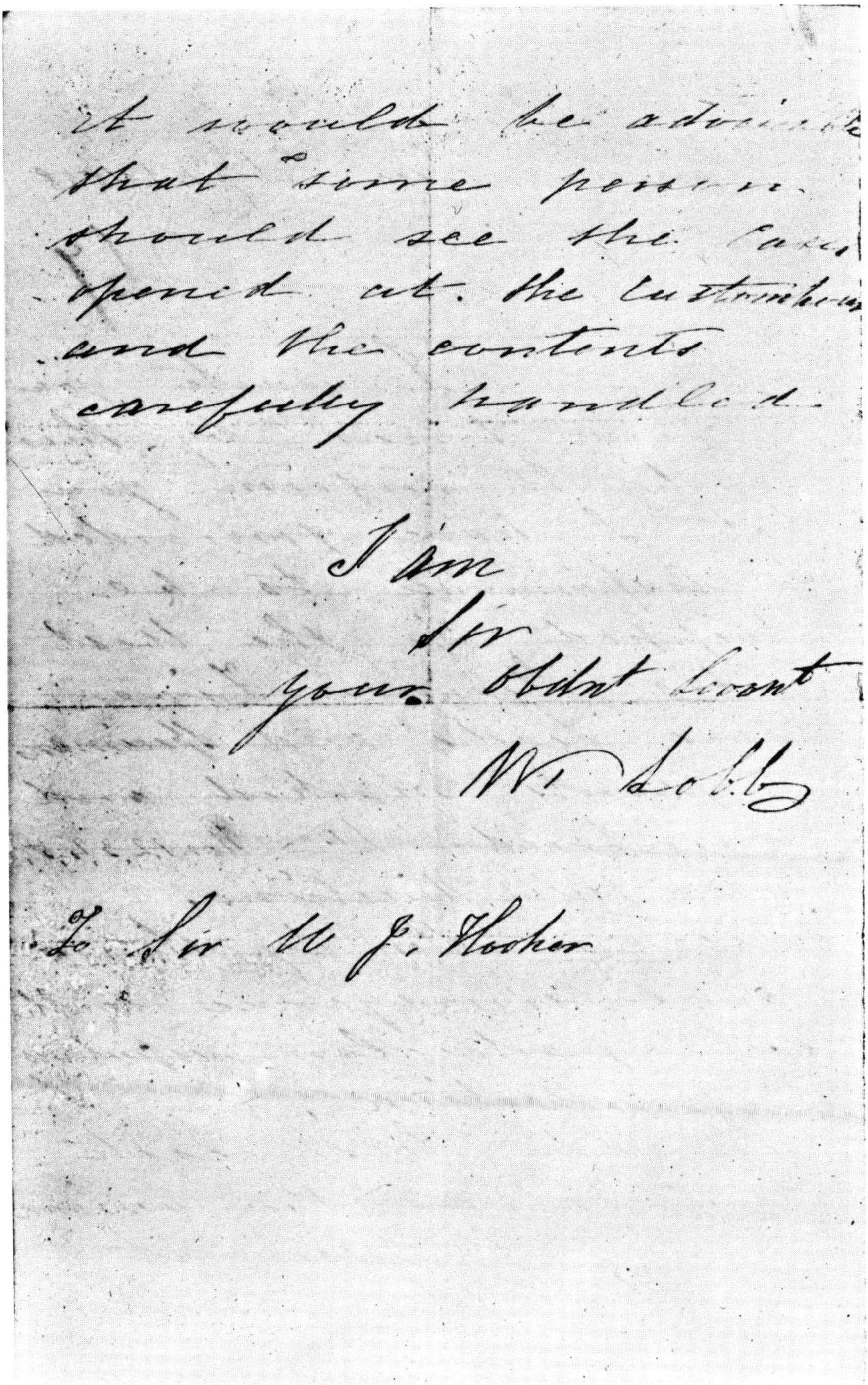

it would be advisable
that some person
should see the Cases
opened at the Custom house
and the contents
carefully handled.

I am
Sir
your obdnt Servnt
W. Lobb

To Sir W. J. Hooker

Headstone formerly in Lone Mountain Cemetery, San Francisco, of William Lobb, photographed 4 October 1936.

ISBN: 0-520-09501-4